The Fourier
Transform
A Tutorial Introduction

James V Stone

The Quantum
Menagerie
A Tutorial Introduction to the
Mathematics of Quantum Mechanics

James V Stone

Artificial
Intelligence
ENGINES
A Tutorial Introduction to the
Mathematics of Deep Learning

James V Stone

Principles of Neural
Information Theory
Computational Neuroscience
and Metabolic Efficiency

James V Stone

Information
Theory
A Tutorial Introduction

James V Stone

Bayes' Rule
A Tutorial Introduction to
Bayesian Analysis
James V Stone

Bayes' Rule
With Python
A Tutorial Introduction to
Bayesian Analysis
James V Stone

Bayes' Rule
With MatLab
A Tutorial Introduction to
Bayesian Analysis
James V Stone

Bayes' Rule
With R
A Tutorial Introduction to
Bayesian Analysis
James V Stone

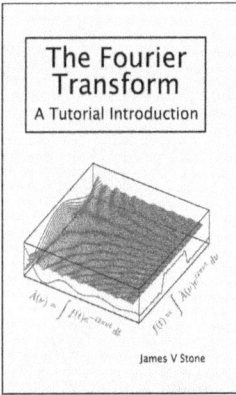

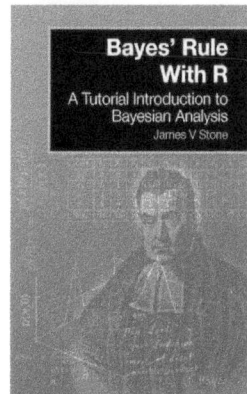

James V Stone, Honorary Associate Professor, Sheffield University, UK.

Linear Regression

A Tutorial Introduction to the
Mathematics of Regression Analysis

James V Stone

Title: Linear Regression
Author: James V Stone

©2022 Sebtel Press

First Edition, 2022.
Typeset in LaTeX $\partial 2_\varepsilon$.
First printing.

ISBN 9781916279193

These are the tears un-cried for you.
So let the oceans
Weep themselves dry.

For Bob, my brother.

Contents

Preface

This book is intended to provide an account of linear regression that is both informal and mathematically rigorous. A large number of diagrams have been included to help readers gain an intuitive understanding of regression, expressed in terms of geometry.

Who Should Read This Book? The material in this book should be accessible to anyone with knowledge of basic mathematics. The tutorial style adopted ensures that readers who are prepared to put in the effort will be rewarded with a solid grasp of regression analysis.

Online Computer Code. Python and Matlab computer code for the numerical example at the end of each chapter can be downloaded from https://github.com/jgvfwstone/Regression.

Corrections. Please send any corrections to j.v.stone@sheffield.ac.uk. A complete list of corrections is available on the book website at https://jim-stone.staff.shef.ac.uk/Regression.

Acknowledgements. Thanks to William Gruner for permission to use his Matlab code for the F and t distributions. Thanks to Nikki Hunkin for valuable feedback, sound advice, and tea. Finally, thanks to Alice Yew for meticulous copyediting and proofreading.

James V Stone.
Sheffield, England.

One of the first things taught in introductory statistics textbooks is that correlation is not causation. It is also one of the first things forgotten.

Thomas Sowell (1930–).

Chapter 1

What is Linear Regression?

1.1. Introduction

Linear regression is the workhorse of data analysis. It is the first step, and often the only step, in fitting a simple model to data. In essence, linear regression is a method for fitting a straight line to a set of data. For example, suppose we have a vague suspicion that tall people have higher salaries than short people. How can we test this hypothesis? Well, we could obtain the salaries and heights of (for example) 13 people, called a *sample*, and plot a graph with salary on one axis and height on the other axis, as in Figure 1.1a. Clearly, the plotted points seem to lie roughly on a straight line, and three plausible lines are shown in Figure 1.1b. But how do we know which (if any) of these lines is the best fitting line? Before exploring this question further, we will summarise the mathematical definition of a line.

1.2. The Equation of a Line

If the relationship between height $\hat{y}$ and salary x is essentially *linear* then it can be described with the equation of a straight line:

$$\hat{y}_i = b_1 x_i + b_0, \tag{1.1}$$

where the *dependent variable* $\hat{y}_i$ is the height of the ith person in the sample, as predicted by the *independent variable* x_i, which is the salary of the ith person in the sample. Already, we can see that a line is defined by two *parameters*, its *slope* b_1 and *intercept* b_0, as shown in Figure 1.2 (the letters b_0 and b_1 are standard notation).

Human height, like most physical quantities, is affected by many factors, so the measured or observed value of a person's height is not perfectly predicted by their salary. By convention, the observed height of the ith individual is denoted by y_i, whereas the height predicted by fitting a straight line to the data is represented as $\hat{y}_i$, as shown in Figure 1.3. The difference between the height predicted by a straight line equation and the observed height is considered to be *noise*, represented by the Greek letter *eta*:

$$\eta_i = y_i - \hat{y}_i. \tag{1.2}$$

Accordingly, the observed height is the predicted height plus noise,

$$y_i = \hat{y}_i + \eta_i. \tag{1.3}$$

For example, the line in Figure 1.2 has a slope of $b_1 = 0.764$ and an intercept of $b_0 = 3.22$, so the predicted value of y_i at x_i is

$$\hat{y}_i = 0.764 x_i + 3.22. \tag{1.4}$$

Deciding which physical variable (salary or height) should be the dependent variable is discussed in Section 1.4.

Slope. The magnitude of the slope b_1 specifies the steepness of the line, and the sign of the slope indicates whether $\hat{y}$ increases or decreases with x. A positive slope means that $\hat{y}$ increases from left to right (as in Figure 1.2), whereas a negative slope means that $\hat{y}$ decreases from left to right.

i	1	2	3	4	5	6	7	8	9	10	11	12	13
x_i	1.00	1.25	1.50	1.75	2.00	2.25	2.50	2.75	3.00	3.25	3.50	3.75	4.00
y_i	3.34	4.97	4.15	5.40	5.21	4.56	3.69	5.86	4.58	6.94	5.57	5.62	6.87

Table 1.1: Values of salary x_i (groats) and measured height y_i (feet) for a fictitious sample of 13 people.

An intuitive understanding of the slope can be gained by considering how a given change in x corresponds to a concomitant change in $\hat{y}$. If we change x from x_1 to x_2 then the change in x is

$$\Delta x \;=\; x_2 - x_1. \tag{1.5}$$

The values of $\hat{y}$ that correspond to x_1 and x_2 are

$$\hat{y}_1 \;=\; b_1 x_1 + b_0, \tag{1.6}$$

$$\hat{y}_2 \;=\; b_1 x_2 + b_0, \tag{1.7}$$

so the change in $\hat{y}$ is

$$\Delta \hat{y} \;=\; \hat{y}_2 - \hat{y}_1, \tag{1.8}$$

which can be rewritten as

$$\Delta \hat{y} \;=\; (b_1 x_2 + b_0) - (b_1 x_1 + b_0). \tag{1.9}$$

The intercept terms b_0 cancel, so

$$\Delta \hat{y} \;=\; b_1(x_2 - x_1) \tag{1.10}$$

$$\;=\; b_1 \Delta x. \tag{1.11}$$

Dividing both sides by Δx yields the slope of the line,

$$b_1 \;=\; \frac{\Delta \hat{y}}{\Delta x}. \tag{1.12}$$

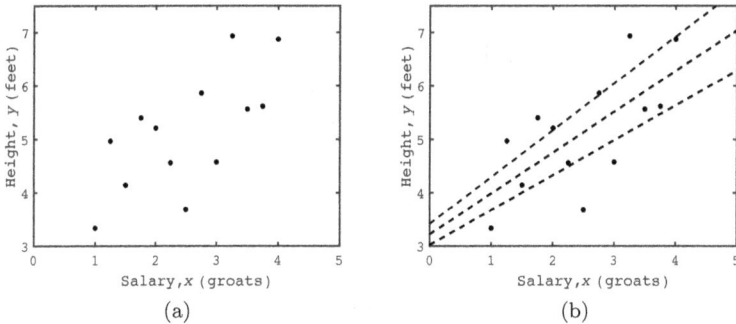

Figure 1.1: (a) Scatter plot of the salary and height data in Table 1.1. (b) Three plausible lines that seem to fit the data. A groat is an obsolete unit of currency, which was worth four pennies in England.

1 What is Linear Regression?

In words, the slope b_1 is the amount of change $\Delta \hat{y}$ in the predicted value $\hat{y}$ of y that corresponds to a change of Δx in x.

This is shown graphically in Figure 1.2, where the triangle has a horizontal length of Δx and a vertical length of $\Delta \hat{y}$. For example, if we choose salary values $x_1 = 1$ and $x_2 = 4$ then the change in salary is

$$\Delta x = x_2 - x_1 \tag{1.13}$$
$$= 4 - 1 \tag{1.14}$$
$$= 3 \text{ groats.} \tag{1.15}$$

From Equation 1.4 the values of $\hat{y}$ at x_1 and x_2 are $\hat{y}_1 = 3.984$ and $\hat{y}_2 = 6.276$ (respectively), so the change in $\hat{y}$ is

$$\Delta \hat{y} = \hat{y}_2 - \hat{y}_1 \tag{1.16}$$
$$= 6.276 - 3.984 \tag{1.17}$$
$$= 2.29 \text{ feet,} \tag{1.18}$$

and therefore the slope of the line is

$$b_1 = \Delta \hat{y}/\Delta x \tag{1.19}$$
$$= 2.29/3 \tag{1.20}$$
$$= 0.764 \text{ feet/groat.} \tag{1.21}$$

Thus, for a salary increase of one groat, height increases by 0.764 feet.

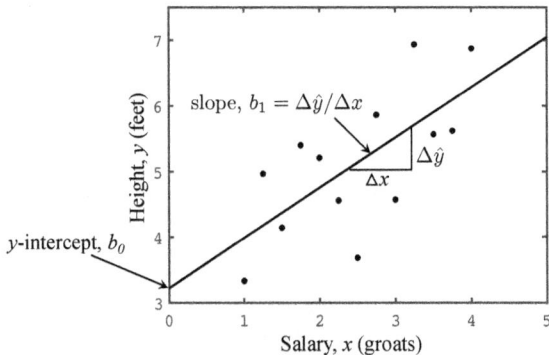

Figure 1.2: A line drawn through the data points. The line's slope is $b_1 = 0.764$ feet/groat, and the intercept at a salary of zero groats is $b_0 = 3.22$ feet, so the equation of the line is $\hat{y} = 0.764\,x + 3.22$.

Intercept. The intercept b_0 specifies the value of $\hat{y}$ where the line meets the *ordinate* (vertical) axis, that is, where the *abscissa* (horizontal) axis is zero, $x = 0$. The value b_0 of the y-intercept can be obtained from Equation 1.1 with $b_1 = 0.764$ by setting $\hat{y}_1 = 3.984$ and $x_1 = 1$:

$$
\begin{aligned}
b_0 &= \hat{y}_1 - b_1 x_1 & (1.22) \\
&= 3.984 - 0.764 \times 1 & (1.23) \\
&= 3.22 \text{ feet.} & (1.24)
\end{aligned}
$$

For our data, this gives the rather odd prediction that someone with a salary of zero groats would have a height of 3.22 feet.

> **Key point**: Given the equation of a straight line, $\hat{y} = b_1 x + b_0$, the slope b_1 is the amount of change in $\hat{y}$ induced by a change of one unit in x, and the intercept b_0 is the value of $\hat{y}$ at $x = 0$ (i.e. where the line meets the ordinate axis).

1.3. The Best Fitting Line

Given the data set of $n = 13$ pairs of x_i and y_i values in Table 1.1, how should we go about finding the best fitting line for those data?

One possibility is to find the line that makes the sum of all *squared* vertical differences between each value of y_i and the line as small as possible, as in Figure 1.3. It may seem arbitrary to look at the sum of squared differences, but this has a sound statistical justification, as we shall see in Chapter 6.

So, if we wish to find the line that minimises the sum of all squared vertical differences then we had better write an equation that describes this quantity. Recall that the vertical distance between an observed value y_i and the value $\hat{y}_i$ predicted by a straight line equation is η_i (Equation 1.2). Using Equations 1.1 and 1.2, if we sum η_i^2 over all n data points then we have the *sum of squared errors*

$$
E = \sum_{i=1}^{n} \left(y_i - (b_1 x_i + b_0) \right)^2. \qquad (1.25)
$$

The values of b_1 and b_0 that minimise this sum of squared errors E are the *least squares estimates* (LSE) of the slope and intercept, respectively. The method described so far is called *simple linear regression*, to distinguish it from *weighted linear regression* (see next page).

> **Key point:** The values of b_1 and b_0 that minimise the sum of squared errors E are the *least squares estimates* (LSE) of the slope and intercept.

Mean Squared Error. A commonly used measure of error is the *mean squared error* (MSE), which is the average squared difference between observed and predicted y values per data point:

$$\overline{E} = \frac{1}{n} \sum_{i=1}^{n} (y_i - \hat{y}_i)^2. \tag{1.26}$$

Because $\overline{E}$ is just E divided by the number of data points n, the values of b_1 and b_0 that minimise E also minimise $\overline{E}$. Be aware that the mean squared error is sometimes defined with $n-1$ rather than n as the denominator in Equation 1.26, so it is important to check which definition is being used in any given context. The philosophy adopted in this book is that the word *mean* should indicate just that (i.e. the mean of n values is the sum divided by n).

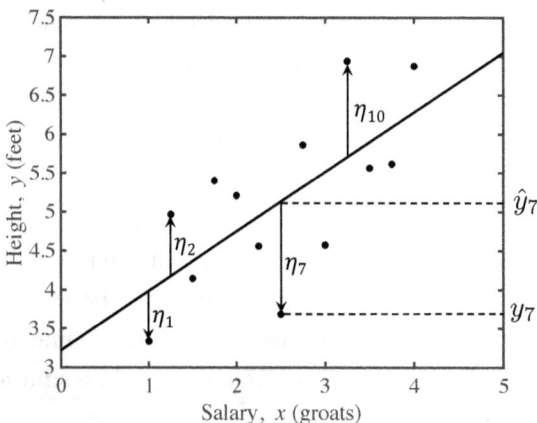

Figure 1.3: The difference η_i between each data point y_i and the corresponding point $\hat{y}_i$ on the best fitting line is assumed to be noise or measurement error. Four examples of $\eta_i = y_i - \hat{y}_i$ for the data in Table 1.1 are shown here.

A Line Suspended on Springs. One way to think about how a line can be adjusted to minimise E is to imagine each difference η_i as a spring that pulls vertically on a free-floating line, so that the line's final position represents the net effect of all those pulling forces. For example, in Figure 1.3, each data point pulls the line along a vertical arrow connecting that point to the line.

However, in practice it is often the case that some observed values are more reliable than others, and the more reliable values should exert greater forces so that they have a greater influence on the final position of the line. Expressed formally, this general strategy yields a method called *weighted linear regression*, which is described in Chapter 8. For now, we assume that all data points are equally reliable.

1.4. Regression and Causation

Before continuing, we should address a question often asked by novices: in our example, why have we decided to define height as the dependent variable and salary as the independent variable, rather than vice versa? After all, to test if taller people have higher salaries, it might seem more natural to think that salary depends on height.

In practice, regression is typically used in experiments to examine the effect of increasing an *independent variable*, or a *regressor*, on the value of a *dependent variable*. In such experiments, the independent variable is controlled by the experimenter and so its value is known exactly. In contrast, values of the dependent variable are the results of physical measurements, which inevitably contain some intrinsic variability or noise. In the data from Table 1.1, the salaries are taken to be values of the independent variable, whereas the heights are taken to be values of the dependent variable. As another example, the variable x could represent the financial rewards for solving different numerical problems in an experiment, and y could represent the amount of time spent trying to solve each problem.

The result of our example regression problem could be interpreted to mean that increasing salary x causes an increase in height y, which is clearly silly. However, it does highlight a common misunderstanding about regression: the fact that we can fit a line to the data in Table 1.1

7

correctly suggests that height increases with increasing salary, but it does not imply that increasing salary *causes* height to increase.

More generally, for any two variables x and y that seem to be *linearly related* (i.e. related by a straight line), it is possible that changes in both x and y are caused by changes in a third variable z. For example, a graph of daily ice cream sales x versus the daily power consumption of air conditioning units y would (plausibly) suggest a straight line relationship. In this case, it is fairly obvious that increasing ice cream sales are not the cause of greater power consumption (or vice versa). Instead, a rise in temperature z is the root cause of increases in both ice cream sales x and power consumption y.

> **Key point**: Fitting a line to a set of data may suggest that one variable y increases with increasing values of another variable x, but it does not imply that increasing x *causes* y to increase.

1.5. Regression: A Summary

Now that we have set the scene, the next four chapters provide details of simple linear regression, summarised here. Given n values of a measured variable y (e.g. height) corresponding to n values of a known quantity x (e.g. salary), find the best fitting line that passes close to this set of n data points. Crucially, the values of y are assumed to contain noise (e.g. measurement error), whereas values of x are assumed to be known exactly. The noise means that the observed values of y do not lie exactly on any single line.

Given that values of y vary, some of this variation can be 'soaked up' (accounted for) by the best fitting line. Precisely how well the best fitting line matches the data is measured as the proportion of the total variation in y that is soaked up by the best fitting line. This proportion can then be translated into a p-value, which is the probability that the slope of the best fitting line is really due to the noise in the measured values of y and that the true slope of the underlying relationship between x and y is actually zero (i.e. a horizontal line).

Later chapters explain regression in relation to maximum likelihood estimation, multivariate regression, weighted linear regression, nonlinear regression, and Bayesian regression.

Chapter 2

Finding the Best Fitting Line

2.1. Introduction

We can estimate the slope b_1 and intercept b_0 of the best fitting line using two different strategies. The first one is *exhaustive search*, which involves trying many different values of b_1 and b_0 in Equation 1.25 to see which values make E as small as possible. Even though exhaustive search is impractical, it gives us an intuitive understanding of how calculus can be used to estimate parameters more efficiently. The second approach is to use calculus to find analytical expressions for the values of b_1 and b_0 that minimise E.

2.2. Exhaustive Search

As mentioned above, one method for finding values of the slope b_1 and intercept b_0 that minimise the sum of squared errors E is to substitute plausible values of b_1 and b_0 into Equation 1.25 (repeated here),

$$E \;=\; \sum_{i=1}^{n} \bigl(y_i - (b_1 x_i + b_0)\bigr)^2, \tag{2.1}$$

and plot the corresponding value of E, as shown in Figure 2.1. The value of E is smallest at $b_1 = 0.764$ and $b_0 = 3.22$. These values are the *least squares estimates* (LSE) of the slope and intercept.

2.3. Onwards and Downwards

Even though the exhaustive search method is effective, it is very labour intensive because it involves evaluating E many times. Fortunately, there is a more efficient method, based on the idea of *gradient descent*.

Gradient descent relies on the intuition that if one wants to get to the bottom of a valley then a simple strategy is to keep moving downhill until there is no more downhill left, at which point the bottom of the valley should have been reached. For our purposes, height in the 'valley' is represented by the sum of squared errors E, and each location on the horizontal ground plane corresponds to a different pair of values of b_1 and b_0, as shown in Figure 2.1.

This is all very well in theory, but if we were standing in this valley, how would we know whether to increase or decrease b_1 and b_0 so as to reduce E? Note that increasing b_1 and b_0 by the same amount corresponds to moving in a north-east direction on the ground plane, whereas decreasing both b_1 and b_0 corresponds to moving in a south-west direction; in fact, every possible combination of changes in b_1 and b_0 corresponds to a particular direction on the ground plane. With this in mind, a simple tactic is to take small steps in different directions on the ground plane, and look for the direction in which the gradient of E points downhill most steeply. Then, having found the direction of steepest descent, a small step in that direction will decrease the

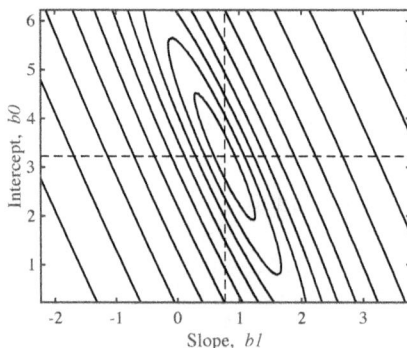

Figure 2.1: Contour map of the values of E for different pairs of values of the slope b_1 and intercept b_0. The point $(b_1, b_0) = (0.764, 3.22)$ where the dashed lines intersect gives the smallest value of E.

height the most, which corresponds to decreasing E the most. Thus, by incrementally changing the values of b_1 and b_0 to decrease E, the least squares estimates of b_1 and b_0 are (eventually) obtained. This is why the method is called gradient descent.

As shown in Figure 2.2a, taking a horizontal cross-section of the sum-of-squares function E in Figure 2.1 yields a curve with a minimum value at $b_1 = 0.764$, where the gradient of E with respect to b_1 is zero. Similarly, Figure 2.2b shows a vertical cross-section of the function E in Figure 2.1, which is a curve with minimum value at $b_0 = 3.22$, where the gradient of E with respect to b_0 is zero.

However, calculus provides a more efficient way of estimating the gradient or *derivative* of E with respect to b_1 and b_0. In fact, for *quadratic error functions* (such as those based on squared differences), we don't have to take a series of small steps at all. Instead, we can jump directly to a point on the ground plane where the gradient is zero, which corresponds to the bottom of the valley.

2.4. The Normal Equations

The fact that the gradient of E with respect to b_1 is zero at $b_1 = 0.764$ provides a vital clue as to how we can estimate the slope b_1 and intercept b_0 without using exhaustive search.

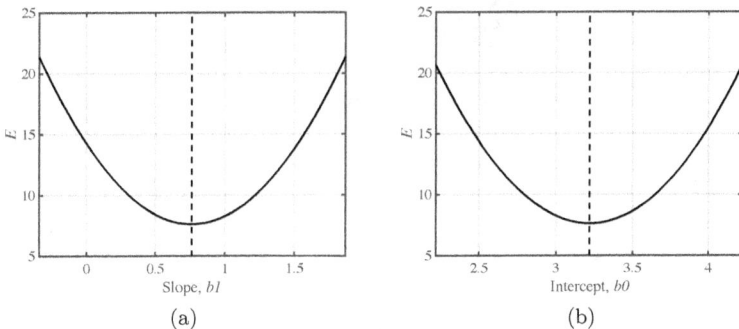

Figure 2.2: (a) Horizontal cross-section of the function E in Figure 2.1 at $b_0 = 3.22$. The value of b_1 that minimises E is $b_1 = 0.764$ (dashed line).
(b) Vertical cross-section of the function E in Figure 2.1 at $b_1 = 0.764$. The value of b_0 that minimises E is $b_0 = 3.22$ (dashed line).

Specifically, taking the derivative of E with respect to b_1 and with respect to b_0 yields a pair of simultaneous equations, the *normal equations*. The solution to the normal equations yields the least squares estimate of b_1 and b_0.

The derivative of E with respect to b_0 is

$$\frac{\partial E}{\partial b_0} = -2\sum_{i=1}^{n}(y_i - (b_1 x_i + b_0)). \tag{2.2}$$

At a minimum of E this equals zero, as shown in Figure 2.2b, so that

$$\sum_{i=1}^{n}(y_i - (b_1 x_i + b_0)) = 0. \tag{2.3}$$

Given that $\eta_i = y_i - (b_1 x_i + b_0)$, if we divide both sides by n we get

$$\bar{\eta} = \frac{1}{n}\sum_{i=1}^{n}(y_i - (b_1 x_i + b_0)) = 0, \tag{2.4}$$

which will prove useful later. Equation 2.3 can be written as

$$\sum_{i=1}^{n} y_i - b_1\sum_{i=1}^{n} x_i - nb_0 = 0, \tag{2.5}$$

which yields the first normal equation,

$$b_1\sum_{i=1}^{n} x_i + nb_0 = \sum_{i=1}^{n} y_i. \tag{2.6}$$

It will prove useful to note that dividing both sides by n yields

$$b_1\bar{x} + b_0 = \bar{y}. \tag{2.7}$$

The derivative of E with respect to b_1 is

$$\frac{\partial E}{\partial b_1} = -2\sum_{i=1}^{n}(y_i - (b_1 x_i + b_0))x_i. \tag{2.8}$$

As shown in Figure 2.2a, at a minimum of E this is equal to zero, which can be written as

$$\sum_{i=1}^{n} x_i y_i - b_1 \sum_{i=1}^{n} x_i^2 - b_0 \sum_{i=1}^{n} x_i = 0, \tag{2.9}$$

giving the second normal equation,

$$b_1 \sum_{i=1}^{n} x_i^2 + b_0 \sum_{i=1}^{n} x_i = \sum_{i=1}^{n} x_i y_i. \tag{2.10}$$

Thus far, we have two simultaneous equations with two unknowns, b_1 and b_0. Equation 2.10 can be used to solve for b_1, as follows. From Equation 2.7 we have $b_0 = \bar{y} - b_1 \bar{x}$, and substituting this into Equation 2.10 gives

$$b_1 \sum_{i=1}^{n} x_i^2 + (\bar{y} - b_1 \bar{x}) \sum_{i=1}^{n} x_i = \sum_{i=1}^{n} x_i y_i. \tag{2.11}$$

Expanding the middle term, we have

$$b_1 \sum_{i=1}^{n} x_i^2 + \bar{y} \sum_{i=1}^{n} x_i - b_1 \bar{x} \sum_{i=1}^{n} x_i = \sum_{i=1}^{n} x_i y_i. \tag{2.12}$$

Collecting the terms containing b_1, we get

$$b_1 \left(\sum_{i=1}^{n} x_i^2 - \bar{x} \sum_{i=1}^{n} x_i \right) = \sum_{i=1}^{n} x_i y_i - \bar{y} \sum_{i=1}^{n} x_i. \tag{2.13}$$

Then, solving for b_1 yields

$$b_1 = \frac{\sum_{i=1}^{n} x_i y_i - \bar{y} \sum_{i=1}^{n} x_i}{\sum_{i=1}^{n} x_i^2 - \bar{x} \sum_{i=1}^{n} x_i}. \tag{2.14}$$

Dividing numerator and denominator by n, we obtain

$$b_1 = \frac{\left(\frac{1}{n} \sum_{i=1}^{n} x_i y_i \right) - \bar{y}\,\bar{x}}{\left(\frac{1}{n} \sum_{i=1}^{n} x_i^2 \right) - \bar{x}^2}. \tag{2.15}$$

Finally, it can be shown that Equation 2.15 can be expressed as

$$b_1 = \frac{\frac{1}{n}\sum_{i=1}^{n}(x_i - \overline{x})(y_i - \overline{y})}{\frac{1}{n}\sum_{i=1}^{n}(x_i - \overline{x})^2}. \tag{2.16}$$

Having found b_1, now the value of b_0 can be obtained from

$$b_0 = \overline{y} - b_1\overline{x}. \tag{2.17}$$

> **Key point:** Taking the derivatives of E with respect to b_1 and b_0 yields a pair of simultaneous equations, the *normal equations*. The solution to the normal equations yields the LSE of b_1 and b_0.

Regressing x On y Versus Regressing y On x

What we have done so far is regressing y on x, i.e. treating y as the dependent variable and x as the independent variable, and seeking a line of the form $\hat{y} = b_1x + b_0$. If we regress x on y, seeking a line of the form $\hat{x} = b_1y + b_0$ as shown in Figure 2.3, we will get different values for the slope and intercept parameters.

Regressing y on x yields a slope and an intercept of

$$b_1 = 0.764 \text{ feet/groat} \quad \text{and} \quad b_0 = 3.22 \text{ feet}, \tag{2.18}$$

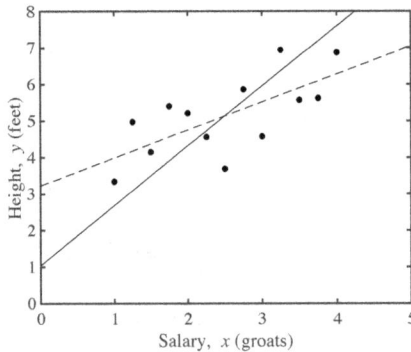

Figure 2.3: Comparison of regressing x on y and regressing y on x. The dashed (best fitting) line for regressing y on x has a slope of $b_1 = 0.764$ feet/groat and minimises the sum of squared *vertical* distances between each data point and the dashed line. The solid (best fitting) line for regressing x on y has a slope of $b_1' = 0.610$ groats/foot and minimises the sum of squared *horizontal* distances between each data point and the solid line.

where $b_1 = \Delta\hat{y}/\Delta x$ feet/groat and b_0 is the value of $\hat{y}$ at $x = 0$.

In contrast, regressing x on y could be achieved by swapping x and y in Equation 1.1 ($\hat{y}_i = b_1 x_i + b_0$) to get

$$\hat{x}_i = b_1' y_i + b_0', \tag{2.19}$$

where the slope $b_1' = \Delta\hat{x}/\Delta y$ and intercept b_0' are

$$b_1' = 0.610 \text{ groats/foot} \quad \text{and} \quad b_0' = -0.635 \text{ groats}. \tag{2.20}$$

The reason that the two best fitting lines in Figure 2.3 are different is that whereas regressing y on x finds the parameter values that minimise the sum of squared (vertical) differences between each observed value of y and the fitting line, regressing x on y finds the parameter values that minimise the sum of squared (horizontal) differences between each value of x and the fitting line. This is discussed in more detail in Section 3.3.

2.5. Numerical Example

Slope. From Equation 2.16,

$$b_1 = 0.6687/0.875 = 0.764 \text{ feet/groat}. \tag{2.21}$$

Intercept. From Equation 2.17 with

$$\bar{x} = 2.50 \quad \text{and} \quad \bar{y} = 5.13, \tag{2.22}$$

we find

$$b_0 = 5.13 - 0.764 \times 2.50 \tag{2.23}$$
$$= 3.225 \text{ feet}. \tag{2.24}$$

Now we know how to find the least squares estimates of the slope b_1 and intercept b_0 of the best fitting line. Next, we consider this crucial question: How well does the best fitting line fit the data?

Chapter 3

How Good is the Best Fitting Line?

3.1. Introduction

Having found the LSE for the slope and intercept of the best fitting line, we naturally wish to get some idea of exactly how well this line actually fits the data. The informal idea of how well a line fits a set of data boils down to how well that line accounts for the variability in the data. In essence, we seek an answer to the following question: for data with a given amount of variability, what proportion of that variability can be explained in terms of the best fitting line? In other words, we wish to know how much of the overall variability in the data can be 'soaked up' by the best fitting line. This requires a formal definition of variability.

3.2. Variance and Standard Deviation

A measure of the variability of n observed values of y is the *variance*,

$$\text{var}(y) \quad = \quad \frac{1}{n}\sum_{i=1}^{n}(y_i - \bar{y})^2. \qquad (3.1)$$

Here are some important properties of the variance.

1. The variance is the mean value of the squared difference between each observed value of y and the mean value $\bar{y}$ of y.

2. If all values of y are multiplied by a constant factor k then the variance increases by a factor of k^2; that is, $\text{var}(ky) = k^2 \, \text{var}(y)$.

3. Adding any constant (e.g. $-\overline{y}$) to all values of y does not change the variance; for example, $\mathrm{var}(y - \overline{y}) = \mathrm{var}(y)$.

4. Variance is related to the sum of squared errors by a factor of n; that is, $\sum(y_i - \overline{y})^2 = n\,\mathrm{var}(y)$.

5. In the context of linear regression, the variance of y can be split into two parts: a) the variance in y that can be predicted from x, and b) the residual or *noise variance* in y that cannot be predicted from x.

A related measure of variability is the *standard deviation*, which is the square root of the variance,

$$s_y \;=\; \sqrt{\mathrm{var}(y)}, \tag{3.2}$$

where s_y is the conventional symbol for the sample standard deviation. If all values of y are multiplied by a constant factor k then the standard deviation increases by a factor of k; that is, $s_{ky} = ks_y$. The standard deviation has the same units as y; for example, if y is measured in units of feet then the standard deviation also has units of feet.

The variance $\mathrm{var}(y)$ is calculated from a finite sample of n values $\{y_1, \ldots, y_n\}$, which is assumed to have been drawn from an infinite *parent population* of y values. Thus, $\mathrm{var}(y)$ is an estimate of the *population variance* σ_y^2, where by convention the Greek letter σ (sigma) is used to represent the parent population standard deviation.

3.3. Covariance and Correlation

Covariance. A useful measure of the degree of association between two variables x and y is their *covariance*,

$$\mathrm{cov}(x,y) \;=\; \frac{1}{n}\sum_{i=1}^{n}(x_i - \overline{x})(y_i - \overline{y}). \tag{3.3}$$

It is worth noting a few important properties of the covariance.

1. A positive covariance indicates that y increases with increasing values of x, and a negative covariance indicates that y decreases with increasing values of x.

2. Adding a constant to x or to y does not change their covariance; for example, $\text{cov}(x - \overline{x}, y - \overline{y}) = \text{cov}(x, y)$.

3. The covariance is sensitive to the magnitudes of x and y. Specifically, if x and y are both multiplied by a factor of k then the covariance increases by a factor of k^2: $\text{cov}(kx, ky) = k^2 \text{cov}(x, y)$.

Like the variance, the covariance $\text{cov}(x, y)$ is based on a finite sample of n values, assumed to be drawn from an infinite (parent) population of values, so $\text{cov}(x, y)$ is an estimate of the *population covariance* $\sigma_{x,y}^2$.

Correlation. Correlation is a measure of association that is independent of the magnitudes of x and y. A commonly used correlation measure is the *Pearson product-moment correlation coefficient,*

$$r \quad = \quad \frac{\frac{1}{n}\sum_{i=1}^{n}(x_i - \overline{x})(y_i - \overline{y})}{\left(\frac{1}{n}\sum_{i=1}^{n}(x_i - \overline{x})^2\right)^{1/2}\left(\frac{1}{n}\sum_{i=1}^{n}(y_i - \overline{y})^2\right)^{1/2}}. \quad (3.4)$$

The $\frac{1}{n}$ terms cancel, but have been left in the expression so that we can recognise it as the covariance (Equation 3.3) divided by the standard deviations of x and y (Equation 3.2):

$$r \quad = \quad \text{cov}(x, y)/(s_x s_y). \quad (3.5)$$

The correlation can vary between $r = \pm 1$. Examples of data sets with different correlations are shown in Figure 3.1.

Back to the Slope. In Equation 2.16 (repeated here),

$$b_1 \quad = \quad \frac{\frac{1}{n}\sum_{i=1}^{n}(x_i - \overline{x})(y_i - \overline{y})}{\frac{1}{n}\sum_{i=1}^{n}(x_i - \overline{x})^2}, \quad (3.6)$$

we can now recognise the numerator as the covariance (Equation 3.3) and the denominator as the variance of x,

$$\text{var}(x) \quad = \quad \frac{1}{n}\sum_{i=1}^{n}(x_i - \overline{x})^2, \quad (3.7)$$

19

so that the slope of the best fitting line is given by the ratio

$$b_1 = \text{cov}(x,y)/\text{var}(x). \qquad (3.8)$$

Therefore, the slope of the best fitting line when regressing y on x is the covariance between x and y expressed in units of the variance of x.

> **Key point:** The slope of the best fitting line when regressing y on x is the covariance between x and y expressed in units of the variance of x.

If we divide all x values by the standard deviation s_x, we obtain a *normalised* variable $x' = x/s_x$, which has a standard deviation of 1, i.e. $s_{x'} = 1$, and hence $\text{var}(x') = (s_{x'})^2 = 1$ as well; similarly, y can be normalised to have standard deviation and variance equal to 1. If both x and y are normalised then the regression coefficient in Equation 3.8 becomes equal to the correlation coefficient in Equation 3.5. In other words, for normalised variables, the slope of the best fitting line equals the correlation, i.e. $r = b_1$.

From Equation 3.5, we have $\text{cov}(x,y) = r s_x s_y$, and substituting this into Equation 3.8 yields $b_1 = r(s_y/s_x)$, so the slope is the correlation scaled by the ratio of standard deviations. Regressing y on x (as above) yields $b_1 = r(s_y/s_x)$, whereas regressing x on y yields $b_1' = r(s_x/s_y)$ (see Figure 2.3).

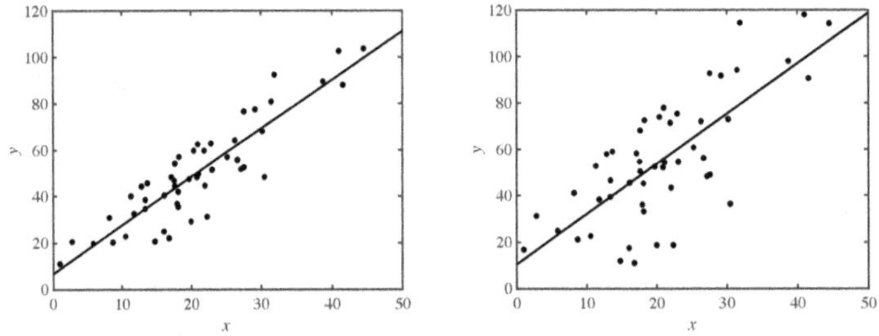

(a)   (b)

Figure 3.1: Comparison of the correlations, covariances and best fitting lines for two data sets.
(a) Correlation $r = 0.90$; $\text{cov}(x,y) = 186$; the best fitting line has a slope of $b_1 = 2.09$ and an intercept of $b_0 = 6.60$.
(b) Correlation $r = 0.75$; $\text{cov}(x,y) = 193$; the best fitting line has a slope of $b_1 = 2.16$ and an intercept of $b_0 = 10.4$.

3.4. Partitioning the Variance

The total error, or difference between each observed value y_i and the mean $\overline{y}$ of y, can be split or *partitioned* into two parts, which we refer to as *signal* and *noise*, as shown in Figure 3.2. The *signal* part of the error is

$$\psi_i \;=\; \hat{y}_i - \overline{y}, \tag{3.9}$$

where $\hat{y}_i = b_1 x_i + b_0$ is the y value corresponding to x_i as predicted by the best fitting line from the regression model. The *noise* part of the error is

$$\eta_i \;=\; y_i - \hat{y}_i, \tag{3.10}$$

which is the part of the error that cannot be explained by the model. The total error of each data point is the sum of the signal and noise

$$y_i - \overline{y} \;=\; (\hat{y}_i - \overline{y}) + (y_i - \hat{y}_i) \tag{3.11}$$

$$\;=\; \psi_i + \eta_i. \tag{3.12}$$

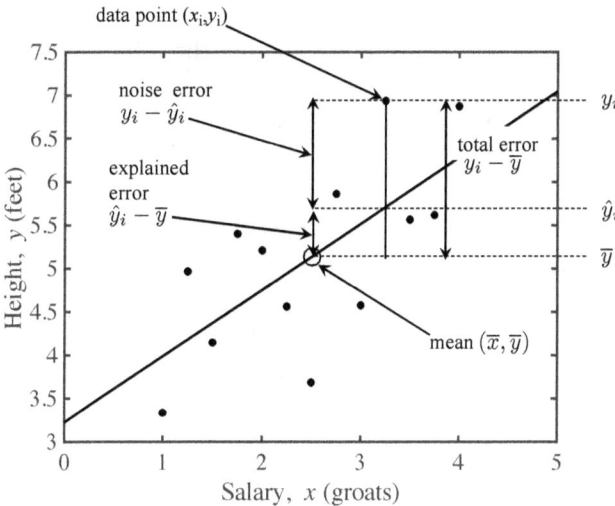

Figure 3.2: The total error (difference between an observed value y_i and the mean $\overline{y}$) can be partitioned into two parts, a signal part, $\hat{y}_i - \overline{y}$, that is explained by the best fitting line and a noise part, $y_i - \hat{y}_i$, which is not explained by the line.

Substituting this into Equation 3.1, the variance can be written as

$$\text{var}(y) \quad = \quad \frac{1}{n}\sum_{i=1}^{n}(\psi_i + \eta_i)^2, \qquad (3.13)$$

which expands to

$$\text{var}(y) \quad = \quad \frac{1}{n}\sum_{i=1}^{n}(\psi_i^2 + \eta_i^2 + 2\psi_i\eta_i). \qquad (3.14)$$

Collecting the terms using separate summation symbols gives

$$\text{var}(y) \quad = \quad \frac{1}{n}\sum_{i=1}^{n}\psi_i^2 + \frac{1}{n}\sum_{i=1}^{n}\eta_i^2 + \frac{2}{n}\sum_{i=1}^{n}\psi_i\eta_i, \qquad (3.15)$$

where the final sum equals zero (see Appendix D, Equation D.21); because (as we should expect) the correlation between signal and noise is zero. Accordingly, the variance can be partitioned into two (signal and noise) parts:

$$\text{var}(y) \quad = \quad \frac{1}{n}\sum_{i=1}^{n}\psi_i^2 + \frac{1}{n}\sum_{i=1}^{n}\eta_i^2. \qquad (3.16)$$

We can multiply both sides by n to obtain three sums of squares,

$$\sum_{i=1}^{n}(y_i - \overline{y})^2 \quad = \quad \sum_{i=1}^{n}\psi_i^2 + \sum_{i=1}^{n}\eta_i^2. \qquad (3.17)$$

For brevity, we define these three sums of squares (reading from left to right in Equation 3.17) as follows.

a) The total sum of squared errors

$$SS_{\text{T}} \quad = \quad \sum_{i=1}^{n}(y_i - \overline{y})^2. \qquad (3.18)$$

b) The signal sum of squares

$$SS_{\text{Exp}} \quad = \quad \sum_{i=1}^{n}\psi^2 = \sum_{i=1}^{n}(\hat{y}_i - \overline{y})^2, \qquad (3.19)$$

which is the part of the sum of squared errors SS_T that is accounted for, or explained, by the regression model.

c) The noise sum of squares

$$SS_{\text{Noise}} \quad = \quad \sum_{i=1}^{n} \eta^2 = \sum_{i=1}^{n} (y_i - \hat{y}_i)^2, \qquad (3.20)$$

which is the part of the sum of squared errors SS_T that is *not* explained by the regression model.

Now Equation 3.17 can be written succinctly as

$$SS_T \quad = \quad SS_{\text{Exp}} + SS_{\text{Noise}}. \qquad (3.21)$$

It will be shown in the next section that the proportion of the sum of squared errors that is explained by the regression model is

$$r^2 \quad = \quad \frac{SS_{\text{Exp}}}{SS_T}, \qquad (3.22)$$

where r^2 is the square of the correlation coefficient defined in Equation 3.4. Finally, given that $SS_{\text{Exp}} = SS_T - SS_{\text{Noise}}$,

$$r^2 \quad = \quad 1 - \frac{SS_{\text{Noise}}}{SS_T}. \qquad (3.23)$$

Notation. Depending on the software used, the total sum of squares is often represented as $SS_T = SST$, the explained sum of squares as

$$SS_{\text{Exp}} \quad = \quad SSR, \qquad (3.24)$$

which stands for *regression sum of squares*, and the noise sum of squares

$$SS_{\text{Noise}} \quad = \quad SSE \text{ or } RSS, \qquad (3.25)$$

meaning the *error sum of squares* or *residual sum of squares*.

Key point: The total sum of squared errors SS_T consists of two parts: SS_{Exp}, the sum of squared errors explained by the best fitting line, and SS_{Noise}, the sum of squared errors that remains unexplained by the best fitting line.

Disregarding the Intercept

The intercept b_0 is not only unimportant in most cases but also makes for unnecessarily complicated algebra. If we can set b_0 to zero then we can ignore it. Accordingly, in Appendix D we prove that b_0 can be set to zero without affecting the slope of the best fitting line, by transforming the means of x and y to zero. This also sets the mean $\bar{\hat{y}}$ of $\hat{y}$ (for points on the best fitting line) to zero, so we have $\bar{\hat{y}} = \bar{x} = \bar{y} = b_0 = 0$.

For the present, the geometric representation of the proof in Figure 3.3 should suffice. This shows that setting the mean of x to zero is equivalent to simply translating the data along the x-axis, and setting the mean of y to zero amounts to translating the data along the y-axis. Crucially, this process of *centring* the data has no effect on the rate at which y varies with respect to x, so the slope of the best fitting line remains unaltered.

For example, after centring, $\bar{x} = \bar{y} = 0$, so Equation 3.6 simpliflies to

$$b_1 = \frac{\sum_{i=1}^{n} x_i\, y_i}{\sum_{i=1}^{n} x_i^2}, \qquad (3.26)$$

where (strictly speaking) x and y should really be written as the zero-mean variables x' and y'.

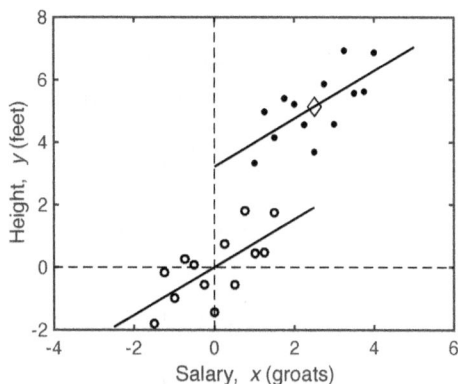

Figure 3.3: Setting the means to zero. The upper right dots represent the original data, with $\bar{x} = 2.50$ and $\bar{y} = 5.13$, marked with a diamond. The lower left circles represent the translated data, with zero means. For the original data (upper right) the best fitting line has equation $\hat{y}_i = b_1 x_i + b_0$, but for the translated data (lower left) $b_0 = 0$ so that $\hat{y}_i = b_1 x_i$. Crucially, the best fitting line has the same slope b_1 for the original and the translated data.

3.5. The Coefficient of Determination

The *coefficient of determination* is the proportion of the variance in y that can be attributed to the best fitting line, i.e. $\mathrm{var}(\hat{y})/\mathrm{var}(y)$. As shown next, this is equal to the square of the correlation coefficient:

$$r^2 = \frac{\mathrm{var}(\hat{y})}{\mathrm{var}(y)}. \tag{3.27}$$

By squaring Equation 3.5, we have

$$r^2 = \frac{\mathrm{cov}(x,y)^2}{\mathrm{var}(x)\mathrm{var}(y)}. \tag{3.28}$$

Given that $\hat{y}_i = b_1 x_i + b_0$, the variance of $\hat{y}$ is

$$\mathrm{var}(\hat{y}) = \frac{1}{n}\sum_{i=1}^{n}(b_1 x_i + b_0 - \bar{\hat{y}})^2. \tag{3.29}$$

By setting $\bar{y} = 0$ we also set $\bar{\hat{y}} = b_0 = 0$ (see page 24), and so

$$\mathrm{var}(\hat{y}) = \frac{1}{n}\sum_{i=1}^{n} b_1^2 x_i^2 \tag{3.30}$$

$$= b_1^2 \,\mathrm{var}(x). \tag{3.31}$$

From Equation 3.8,

$$b_1^2 = \frac{\mathrm{cov}(x,y)^2}{\mathrm{var}(x)^2}. \tag{3.32}$$

Substituting this into Equation 3.31 gives

$$\mathrm{var}(\hat{y}) = \frac{\mathrm{cov}(x,y)^2}{\mathrm{var}(x)} \tag{3.33}$$

and therefore

$$\mathrm{cov}(x,y)^2 = \mathrm{var}(\hat{y})\,\mathrm{var}(x). \tag{3.34}$$

Substituting this in Equation 3.28 yields Equation 3.27, as promised.

Similarly, Equation 3.22 is obtained by first writing Equation 3.27 as

$$r^2 = \frac{\frac{1}{n}\sum_{i=1}^{n}(\hat{y}_i - \overline{\hat{y}})^2}{\frac{1}{n}\sum_{i=1}^{n}(y_i - \overline{y})^2}. \tag{3.35}$$

Then, by setting $\overline{y} = 0$ so that $\overline{\hat{y}} = b_0 = 0$ (see page 24), we obtain

$$r^2 = \frac{\sum_{i=1}^{n}\hat{y}_i^2}{\sum_{i=1}^{n}y_i^2} \tag{3.36}$$

$$= \frac{SS_{\text{Exp}}}{SS_{\text{T}}}, \tag{3.37}$$

which proves Equation 3.22.

> **Key point**: The coefficient of determination, the proportion of variance that can be attributed to the best fitting line, is equal to the square of the correlation coefficient, r^2.

3.6. Numerical Example

From the data in Table 1.1, we have

$$\text{var}(y) = 1.095 \quad \text{and} \quad \text{var}(\hat{y}) = 0.511 \tag{3.38}$$

Hence, the proportion of the overall variance accounted for by the best fitting line is (using Equation 3.27)

$$r^2 = \text{var}(\hat{y})/\text{var}(y) \tag{3.39}$$

$$= 0.511/1.095 = 0.466. \tag{3.40}$$

So just under half of the total variance in y can be attributed to the best fitting line, and just over half is simply noise. The correlation between x and y is $r = \sqrt{0.466} = 0.683$.

Two further examples can be seen in Figure 3.1, where it is apparent that data which lie close to the best fitting line allow a better fit to that line. In the next chapter, we consider how to assess the statistical significance of mean values, which paves the way for assessing the statistical significance associated with the coefficient of determination r^2.

Chapter 4

Statistical Significance: Means

4.1. Introduction

Having estimated the slope of the best fitting line, how can we find the *statistical significance* associated with that slope?

Just as the best fitting slope is an estimate of the true slope of the relationship between two variables, so the mean of n values in a sample is an estimate of the mean of the population from which the sample was drawn. It turns out that the estimated slope is a weighted mean (as will be shown in Section 5.2), which is a generalisation of a conventional mean. Therefore, we can employ standard methods for finding the statistical significance of a mean value to find the statistical significance of the best fitting slope.

To calculate a conventional mean, we add up all n measured quantities and divide the sum by n. For a weighted mean, each measured quantity is boosted or diminished according to the value of its associated weight, so some measurements contribute more to the weighted mean. The justification is that some measured quantities are more reliable than others, and such values should contribute more to the weighted mean. Accordingly, and just for this chapter, we will treat the estimated slope as if it were a conventional mean (i.e. an unweighted mean).

4.2. The Distribution of Means

The data set in Table 1.1 is just one of many possible sets of data, or samples, taken from an underlying large collection of salary and height values. Indeed, we can treat our data set as if it is a single sample of values drawn from an infinitely large *parent population* of values. If the *population mean* is μ then we can treat each observed value of y as if it consists of two components, μ plus random noise ε:

$$y_i = \mu + \varepsilon_i. \tag{4.1}$$

To get an idea of how typical our sample is, suppose we could take many samples. Consider a random sample of n values, $V_1 = \{y_1, y_2, \ldots, y_n\}$, where the subscript 1 indicates that this is the first sample; we denote its mean value by $\bar{y}_1$. If we repeat this sampling procedure N times, we get N samples $\{V_1, V_2, \ldots, V_N\}$, where each sample V_j has mean $\bar{y}_j$, so we have N means,

$$\{\bar{y}_1, \bar{y}_2, \ldots, \bar{y}_N\}. \tag{4.2}$$

If we make a histogram of these N mean values then we will find that they cluster around the mean μ of the parent population, as in Figure 4.1. The mean value of the jth sample of n values is

$$\bar{y}_j = \frac{1}{n}\sum_{i=1}^{n} y_{ij} = \frac{1}{n}\sum_{i=1}^{n}(\mu + \varepsilon_{ij}), \tag{4.3}$$

where y_{ij} denotes the ith value of y in the jth sample and ε_{ij} is the noise in y_{ij}. By splitting this into two summations, we have

$$\bar{y}_j = \frac{1}{n}\sum_{i=1}^{n}\mu + \frac{1}{n}\sum_{i=1}^{n}\varepsilon_{ij}. \tag{4.4}$$

The first term is just μ, and the second term is the mean of the noise values in the jth sample,

$$\bar{\varepsilon}_j = \frac{1}{n}\sum_{i=1}^{n}\varepsilon_{ij}, \tag{4.5}$$

so Equation 4.4 becomes

$$\bar{y}_j = \mu + \bar{\varepsilon}_j. \tag{4.6}$$

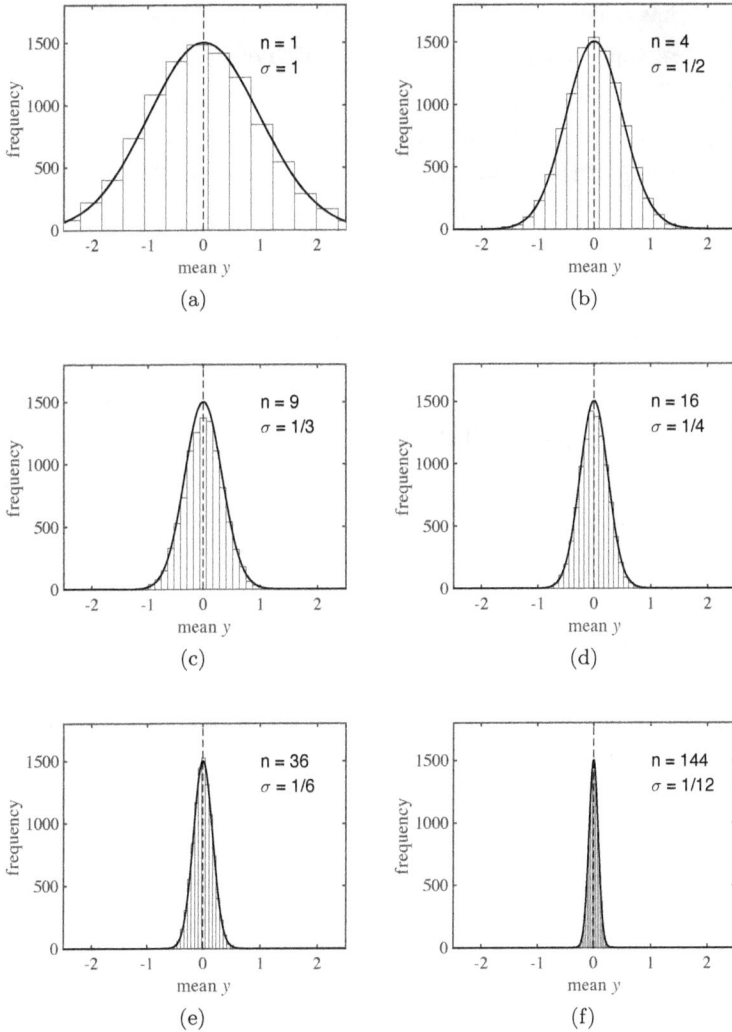

Figure 4.1: Each panel shows a histogram of $N = 10{,}000$ means. All the means in each histogram are based on a single sample size n, and n increases from (a) to (f), as indicated in each panel. For example, in (b), first the mean $\bar{y}$ of $n = 4$ randomly chosen values of y was found; then this was repeated to obtain a total of 10,000 means, and a histogram of those means was plotted. Each value of y was chosen from a parent population with mean $\mu = 0$ and standard deviation $\sigma_y = 1$, shown in (a). As n increases, the standard error $\sigma_{\bar{y}}$ shrinks (standard deviation of the n mean values, denoted by σ in each panel). The smooth curve in each panel is a Gaussian function with standard deviation $\sigma_{\bar{y}} = \sigma_y / \sqrt{n}$.

29

The *law of large numbers* says that if the distribution of noise ε in the parent population has a mean $\bar{\varepsilon}$ of zero then the distribution of noise means $\bar{\varepsilon}_j$ has a mean that converges to zero as n increases. Therefore, as n increases, Equation 4.6 becomes $\bar{y} \approx \mu$, with equality as n approaches infinity. Thus, the sample mean $\bar{y}$ is essentially a noisy estimate of the population mean μ. But how much confidence can we place in our estimate $\bar{y}$ of the population mean μ? To answer this, we need to find the standard deviation of the estimated mean $\bar{y}$. And, to find the standard deviation of the mean $\bar{y}$, we need the *central limit theorem*.

The Central Limit Theorem

The central limit theorem states that a histogram of N sample means as in Equation 4.2 has a bell-shaped distribution, known as a *Gaussian distribution*; see Figure 4.2. This theorem is important because, remarkably, the distribution of means is approximately Gaussian *almost irrespective of the shape of the parent distribution of y values*. Incidentally, a theorem is simply a mathematical statement that has been proved to be true.

Given a parent population with standard deviation σ_y, suppose we take a large number of samples that each contain n values. The central limit theorem states that the standard deviation of the sample means is

$$\sigma_{\bar{y}} = \sigma_y/\sqrt{n}. \qquad (4.7)$$

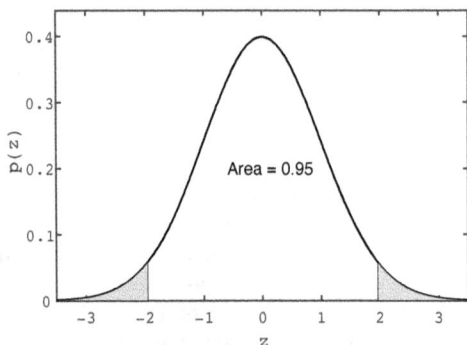

Figure 4.2: A normalised Gaussian distribution has mean $\mu = 0$ and standard deviation $\sigma = 1$. The total area under the curve is 1.0 and the area between $z = \pm 1.96$ is 0.95. For any Gaussian distribution, the area under the curve between ± 1.96 standard deviations is 95% of the total area.

This standard deviation of sample means has a special name, the *standard error*. Given N sample means $\bar{y}_j$ as in Equation 4.2, the estimated value of $\sigma_{\bar{y}}$ is

$$s_{\bar{y}} = \left(\frac{1}{N} \sum_{j=1}^{N} (\bar{y}_j - \mu)^2 \right)^{1/2}. \tag{4.8}$$

The factor of $1/\sqrt{n}$ in Equation 4.7 implies that the standard error shrinks as the sample size n increases. However, there are 'diminishing returns' on increasing n: initially the standard error shrinks rapidly, but for sample sizes above 20 the rate of decrease slows considerably, as shown in Figure 4.3. Despite these diminishing returns, Equation 4.7 guarantees that $\sigma_{\bar{y}}$ approaches zero as n tends to infinity.

The Gaussian (Normal) Distribution. The equation for the probability of a variable $\bar{y}$ with a Gaussian distribution that has mean μ and standard deviation $\sigma_{\bar{y}}$ is

$$p(\bar{y}) = k e^{-(\bar{y}-\mu)^2/(2\sigma_{\bar{y}}^2)}, \tag{4.9}$$

where $e = 2.718\ldots$ and $k = [1/(2\pi\sigma_{\bar{y}}^2)]^{1/2}$, which ensures that the area under the curve sums to 1.

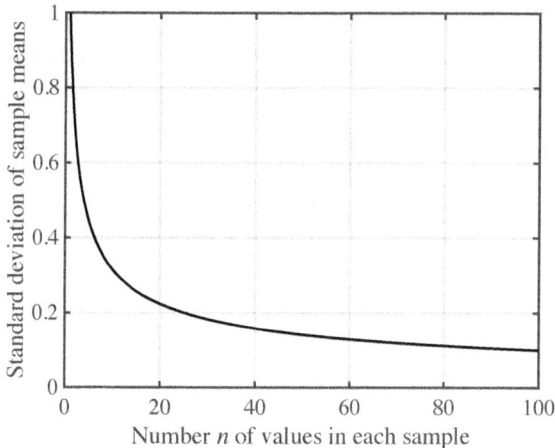

Figure 4.3: Consider a population with a standard deviation of $\sigma_y = 1$. Given N samples from this population, each of which contains n values, the standard deviation of the sample means (standard error) is $\sigma_{\bar{y}} = \sigma_y/\sqrt{n}$ (Equation 4.7).

> **Key point**: Given N samples, each of which consists of n values chosen from a parent population with mean μ and standard deviation σ_y, the distribution of the sample means $\{\overline{y}_1, \ldots, \overline{y}_N\}$ is Gaussian with mean μ and standard deviation $\sigma_{\overline{y}} = \sigma_y/\sqrt{n}$.

4.3. Degrees of Freedom

In essence, the number of degrees of freedom is the number of ways in which a sample of n values are free to vary, after one or more constraints have been taken into account. By convention, the Greek letter ν (pronounced *new*) is standard notation for degrees of freedom. For example, if there are n values in a sample then the number of degrees of freedom is $\nu = n$, because each value is free to vary. But if we wish to calculate the variance, for example, then its associated number of degrees of freedom is $\nu = n - 1$, as explained next.

Variance is calculated from n squared differences $(y_i - \overline{y})^2$, as in Equation 3.1. For a sample of $n = 2$ values, where $y_1 = 6$ and $y_2 = 4$, the sum of both values is $S = 10$ and the mean is $\overline{y} = 5$. If we make a two-dimensional graph with coordinate axes y_1 and y_2 then the values in this sample can be represented as a single point located at $(6, 4)$ on the (y_1, y_2) plane, as shown in Figure 4.4a. When there are no constraints (e.g. if we allow the mean to take any value), y_1 and y_2 can adopt any values, so the point (y_1, y_2) can be anywhere on the plane. But when we are calculating the variance based on a fixed value of the mean $\overline{y}$, the value of the sum $S = n\overline{y}$ is effectively fixed, at 10 in this example. In terms of geometry, fixing the value of the sum S means that the point (y_1, y_2) must lie on a line given by the equation $y_1 + y_2 = S$. In other words, for a sample with a fixed mean (or equivalently a fixed sum) the only combinations of y_1 and y_2 allowed must correspond to points that lie on the line defined by $y_1 + y_2 = S$. Note that a line has one dimension less than the plane defined by the axes y_1 and y_2.

What has this to do with variance? Well (as mentioned above), variance depends on n squared differences from the mean, which is fixed in the calculation. So once the value of y_1 is known, the value of y_2 is also known (it is $y_2 = S - y_1$); similarly, once the value of y_2 is known, the value of y_1 is also known. Therefore, for a sample of $n = 2$ values with a fixed mean $\bar{y}$, only $1 = n - 1$ of the values is free to vary. In effect, the variance confines the sample values (y_1, y_2) to a one-dimensional space (e.g. a line $y_1 + y_2 = S$), so its associated number of degrees of freedom is $\nu = 1$. (If we choose to take account of the fact that the differences are squared then the one-dimensional space is actually a circle, but this has the same number of dimensions (one) as a line, hence the underlying logic of the argument presented here holds good).

By analogy, if the sample comprises $n = 3$ values $\{y_1, y_2, y_3\}$ then this defines a point in three-dimensional space with coordinate axes y_1, y_2 and y_3, as shown in Figure 4.4b. When the mean, and hence the sum S, is fixed, if y_1 and y_2 are known then y_3 can be calculated as $y_3 = S - (y_1 + y_2)$. In other words, once the mean is fixed, the point (y_1, y_2, y_3) has components such that $y_1 + y_2 + y_3 = S$, which defines a plane in three-dimensional space. Because the variance effectively confines the sample values (y_1, y_2, y_3) within this plane, its associated number of degrees of freedom is $\nu = 2$.

Generalising, if the sample comprises n values $\{y_1, \ldots, y_n\}$ then this defines a point in an n-dimensional space with coordinate axes $y_1, \ldots, y_n$. Given a fixed mean and hence a fixed sum S, if $y_1, \ldots, y_{n-1}$ are known

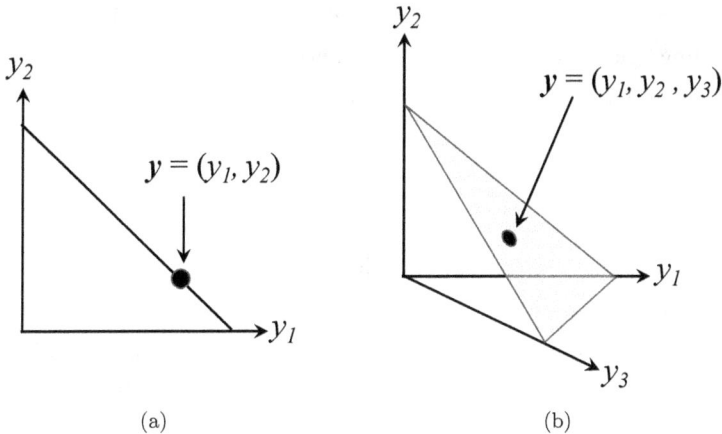

(a)   (b)

Figure 4.4: a) For a sample with $n = 2$ values y_1 and y_2, if their sum S is fixed then the point $\boldsymbol{y} = (y_1, y_2)$ must lie on the line shown, which has equation $y_1 + y_2 = S$. (b) For a sample with $n = 3$ values y_1, y_2 and y_3, if their sum S is fixed then the point $\boldsymbol{y} = (y_1, y_2, y_3)$ must lie on the plane shown, which has equation $y_1 + y_2 + y_3 = S$. One possible value for $\boldsymbol{y}$ is shown in (a) and (b).

then y_n can be calculated as $y_n = S - (y_1 + \cdots + y_{n-1})$. So once the mean is fixed, the point $(y_1, \ldots, y_n)$ has components such that $y_1 + \cdots + y_n = S$, which defines an $(n-1)$-dimensional hyperplane in n-dimensional space. Because the variance effectively confines the sample values $(y_1, \ldots, y_n)$ to this $(n-1)$-dimensional hyperplane, its associated number of degrees of freedom is $\nu = n - 1$.

As a general rule, if a model has p parameters then each parameter acts as a constraint on the degrees of freedom in that model. Consequently, given n data points, a model with p parameters has $\nu = n - p$ degrees of freedom. In the case of simple linear regression, there are $p = 2$ parameters (slope and intercept), so the number of degrees of freedom in the model is $\nu = n - 2$. See Walker (1940) for more details.

4.4. Estimating Variance

It is tempting to assume that the variance var(y) of a sample of n values provides a reasonable estimate of the variance σ_y^2 of the parent population. However, var(y) is a *biased* estimate; specifically, var(y) systematically under-estimates the variance of the parent population.

An unbiased estimate of the population variance σ^2 is obtained by dividing the sum of squared differences by the degrees of freedom $\nu = n - 1$ rather than by the sample size n:

$$\hat{\sigma}_y^2 = \frac{1}{n-1} \sum_{i=1}^{n} (y_i - \bar{y})^2. \tag{4.10}$$

By implication, an unbiased estimate of the standard deviation σ_y of the parent population is $\hat{\sigma}_y$. As a reminder, a variable with a hat (such as $\hat{\sigma}_y$) is an estimate of the hatless quantity (σ_y). Note that the difference between the sum of squared differences divided by n and divided by $n - 1$ becomes negligible as n increases. By analogy with Equation 4.7, the unbiased estimate $\hat{\sigma}_{\bar{y}}$ of the standard error $\sigma_{\bar{y}}$ (i.e. the standard deviation of the sample mean $\bar{y}$) is

$$\hat{\sigma}_{\bar{y}} = \hat{\sigma}_y / \sqrt{n}. \tag{4.11}$$

And the central limit theorem guarantees that the distribution of sample means is approximately Gaussian with a mean of μ and a standard deviation of $\sigma_{\overline{y}}$.

The z-score. If for each sample j we subtract the population mean μ from the sample mean $\overline{y}_j$ and divide by the standard deviation of the mean (standard error) $\sigma_{\overline{y}}$, the result is a *z-score*,

$$z \;=\; (\overline{y} - \mu)/\sigma_{\overline{y}}, \tag{4.12}$$

where $\sigma_{\overline{y}} = \sigma_y/\sqrt{n}$ (Equation 4.7), so that

$$z \;=\; \frac{\overline{y} - \mu}{\sigma_y/\sqrt{n}}. \tag{4.13}$$

The Normalised Gaussian Distribution. A z-score has a *normalised Gaussian distribution*, that is, a Gaussian distribution with a mean of 0 and a standard deviation of 1. As the total area under the curve of a Gaussian distribution is equal to 1 (see page 31), we can interpret area under a portion of the curve as probability, as shown in Figure 4.5. The normalisation makes calculations much simpler.

Why is Noise Gaussian? The justification for assuming that noise is Gaussian is that it usually consists of a mixture of many different variables. Each of these variables can have a unique amplitude, so noise is effectively a weighted sum. Because every weighted sum is proportional to its weighted mean, weighted sums and weighted means have the same distribution. Given that the central limit theorem guarantees that all weighted means are approximately Gaussian, it follows that noise has a Gaussian distribution. For the sake of brevity, in the following we will assume that the distribution of means is Gaussian.

4.5. The p-Value

Because all sets of data contain noise, which has a Gaussian distribution, the central limit theorem tells us that finding the mean of almost any data set is analogous to choosing a point under a Gaussian distribution curve at random. As illustrated in Figure 4.5, the probability of choosing that point is proportional to the height $p(z)$ of the curve above that point. Consequently, the probability of choosing a point located at z along the abscissa decreases with distance from the mean μ. In other words, the Gaussian curve defines the probability of any given value of z, and this probability is a Gaussian function of z as in Equation 4.9.

If we now consider the normalised Gaussian distribution shown in Figure 4.2, which has mean 0 and standard deviation 1, then the area under the curve between ± 1.96 accounts for 95% of the total area, so the probability of choosing a point z that lies between ± 1.96 is 0.95. It follows that the total area of the two 'tails' of the distribution outside of this central region is $p = 1 - 0.95 = 0.05$ (see Figure 4.2). Specifically, the tail of the distribution to the right of $z = +1.96$ has an area of 0.025, and the tail of the distribution to the left of $z = -1.96$ also has an area of 0.025. This means that the probability that z is more than 1.96 is 0.025, and the probability that z is less than -1.96 is 0.025. The probability that the absolute value $|z|$ of z is larger than 1.96 is the p-value, $p = 1 - 0.95 = 0.05$.

Key point: The area under the normalised Gaussian curve between ± 1.96 occupies 95% of the total area. Therefore, the combined probability of the two tails of the distribution beyond this central region is $1 - 0.95 = 0.05$.

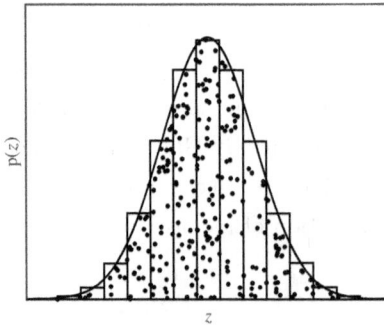

Figure 4.5: A normalised histogram as an approximation to a Gaussian distribution (smooth curve). If dots fall onto the page at random positions, there will usually be more dots in the taller columns (only dots that fall under the curve are shown). The proportion of dots in each column is proportional to the height $p(z)$ of that column. Therefore, the probability of choosing a specific value of z is proportional to $p(z)$.

4.6. The Null Hypothesis

Conventional statistical analysis is based on the idea that data can be used to reject a *null hypothesis*, which usually assumes there is no relationship between variables or no effect of one variable on another. If the data provide sufficient evidence to reject the null hypothesis then an *alternative hypothesis* can be accepted. We will not discuss this in detail, except to say that there may be many alternative hypotheses which could account for the observed data.

As a simple example, consider the sample of 13 values of y in Table 1.1, which has a mean of $\overline{y} = 5.135$. Is this mean significantly different from zero? In other words, what is the probability of obtaining a sample with this mean if the population mean μ is zero?

As will be shown below, it is improbable that the y data in Table 1.1 could have been obtained from a population of values for which the population mean μ equals zero. Precisely how improbable is indicated by the infamous p-value. In this case, we take the null hypothesis to be that the population mean is zero, and we want to see if this hypothesis can be rejected in favour of the alternative hypothesis that the population mean is not zero.

Key point: Given a set of data with nonzero mean, we test whether it is improbable that the mean could have been obtained by pure chance from a population with a mean of zero. Precisely how improbable is indicated by a p-value.

One-Tailed Tests. If we assume that, like the original scenario in Table 1.1, our 13 values of y are measurements of height, there is no point in considering negative values of y. Consequently, we do not consider the possibility of $\mu < 0$, so the alternative hypothesis is that $\mu > 0$ and the null hypothesis is that $\mu = 0$. By convention, any value that occurs with probability less than 0.05 is deemed statistically significant (although p-values of 0.01 or less are used in some research fields). Therefore, the question is: given a distribution with mean zero, which values of the sample mean $\overline{y}$ are big enough to occur with probability less than (or equal to) 0.05?

The answer is: any value of $\bar{y}$ that lies under the shaded area of the Gaussian distribution in Figure 4.6a. Because the total area under the distribution curve is 1, the area of any region under the curve corresponds to a probability. Imagine increasing the value of $\bar{y}$ in Figure 4.6a and calculating the area under the curve to the right of the current value; it turns out that when the area under the curve to the right of $\bar{y}$ equals 0.05, the location of $\bar{y}$ corresponds to a critical value of $\bar{y}_{\text{crit}} = 1.645 \times \sigma_{\bar{y}}$, where $\sigma_{\bar{y}}$ is the standard error. In other words, the value of $\bar{y}$ that is 1.645 standard errors above the mean cuts off a shaded region with an area of 0.05.

Using Equation 4.11, the estimated standard error is $\hat{\sigma}_{\bar{y}} = 0.302$, so $\bar{y}_{\text{crit}} = 1.645 \times 0.302 = 0.497$. Therefore, values of $\bar{y}$ larger than 0.497 are considered to be statistically significantly different from zero. Clearly, the observed mean of 5.135 is larger than 0.497, so we can reject the null hypothesis that $\mu = 0$. To put it another way, the observed mean of $\bar{y} = 5.135$ is $5.135/0.302 = 17.003$ standard errors above zero, which is very significant. Because we only consider the area in the tail on one side of the distribution, this is called a one-tailed test.

Two-Tailed Tests. Now suppose that the 13 values of y in Table 1.1 are temperature measurements, so the mean temperature of this sample

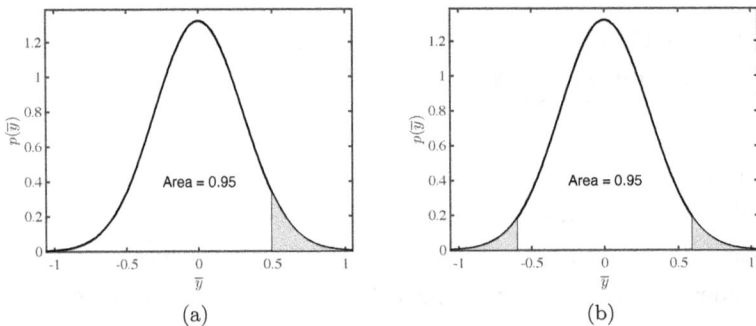

(a)   (b)

Figure 4.6: Two views of a Gaussian distribution with mean $\mu = 0$ and standard deviation $\sigma_{\bar{y}} = 0.302$. (a) The shaded region in the right tail has an area of 0.05; the probability that a sample of n values has a mean at least as large as 0.497 is equal to the area of the shaded region (0.05). (b) Each of the two shaded regions has an area of 0.025, so the total shaded area is 0.05; the probability that a sample of n values has a mean of 0.592 or greater is 0.025, and the probability that the mean is -0.592 or more negative is also 0.025.

is $\overline{y} = 5.135$ degrees. In this case, since temperatures can be negative, we must consider the possibility that $\overline{y}$ and μ could be less than zero. As in the one-tailed test, the null hypothesis is that $\mu = 0$, but now the alternative hypothesis is not just that $\mu > 0$; instead, it is a composite hypothesis: either $\mu > 0$ or $\mu < 0$. Equivalently, the alternative hypothesis is that $\mu \neq 0$.

We calculated that the estimated standard deviation of the sample mean is $\hat{\sigma}_{\overline{y}} = 0.302$, so we can work out that 2.5% of the total area under the curve lies to the right of the critical value $\overline{y}_{\text{crit}}^{+} = 1.96 \times \sigma_{\overline{y}} = 1.96 \times 0.302 = 0.592$; similarly, 2.5% of the total area under the curve lies to the left of the critical value $\overline{y}_{\text{crit}}^{-} = -0.592$. In other words, the area to the right of 0.592 is 0.025, and the area to the left of -0.592 is also 0.025, as shown in Figure 4.6b. Therefore, the probability that $|\overline{y}| > 0.592$ is less than 0.05. So if $\overline{y} > 0.592$ then the p-value is $p < 0.05$ (i.e. a statistically significant p-value). In fact, our sample mean is $\overline{y} = 5.135$, which is larger than $\overline{y}_{\text{crit}}^{+}$, so we can reject the null hypothesis (i.e. we can conclude that μ is significantly different from zero). Because we consider the areas in both the left and the right tails of the distribution, this is called a two-tailed test.

4.7. The z-Test

If the population standard deviation σ_y is known then we can test whether a sample mean $\overline{y}$ is significantly different from zero by using the z-score in Equation 4.12; this is called the z-test. In practice, σ_y is never known, but assuming it to be known allows us to explain some basic principles that will be used in the next section.

The standard deviation of the mean (standard error) $\sigma_{\overline{y}}$ can be obtained from the population standard deviation σ_y using $\sigma_{\overline{y}} = \sigma_y/\sqrt{n}$ (Equation 4.7). As $z = (\overline{y} - \mu)/\sigma_{\overline{y}}$ in Equation 4.12 has a Gaussian distribution with mean 0 and standard deviation 1 (see page 35 and Figure 4.5), there is a 95% chance that z lies between ± 1.96,

$$-1.96 \ \leq \ z \ \leq \ +1.96, \tag{4.14}$$

that is,

$$-1.96 \quad \leq \quad (\bar{y} - \mu)/\sigma_{\bar{y}} \quad \leq \quad +1.96. \tag{4.15}$$

Equivalently, there is a 95% chance that

$$\mu - 1.96\,\sigma_{\bar{y}} \quad \leq \quad \bar{y} \quad \leq \quad \mu + 1.96\,\sigma_{\bar{y}}, \tag{4.16}$$

which means there is a 95% chance that the sample mean $\bar{y}$ is within 1.96 standard errors of the population mean μ. More importantly, Equation 4.15 can be rewritten as

$$\bar{y} - 1.96\,\sigma_{\bar{y}} \quad \leq \quad \mu \quad \leq \quad \bar{y} + 1.96\,\sigma_{\bar{y}}, \tag{4.17}$$

which means there is a 95% chance that the population mean μ lies between ± 1.96 standard errors of the sample mean $\bar{y}$.

Specifically, there is a 95% chance that the population mean μ lies in the *confidence interval* (CI) between the *confidence limits* $\bar{y} - 1.96\,\sigma_{\bar{y}}$ and $\bar{y} + 1.96\,\sigma_{\bar{y}}$, which is written as

$$\mathrm{CI}(\mu) \quad = \quad \bar{y} \pm 1.96\,\sigma_{\bar{y}}. \tag{4.18}$$

For a Gaussian distribution, 95% of the area under the curve is within ± 1.96 standard deviations of the mean. This implies that each of the two tails of the distribution contains 2.5% of the area, making a total of 5%. Accordingly, $z_{0.05} = 1.96$ is the *critical value* of z for a *statistical significance level* or *p-value* of 0.05.

4.8. The *t*-Test

Given that the population standard deviation σ_y, and hence the standard error $\sigma_{\bar{y}}$, is not known, testing whether an observed mean $\bar{y}$ is significantly different from zero requires a *t*-test.

If we replace the unknown standard error $\sigma_{\bar{y}}$ in Equation 4.12 with its unbiased estimate $\hat{\sigma}_{\bar{y}}$ from Equations 4.11 and 4.10 then the distribution of the resulting variable will not quite be Gaussian. In fact, the distribution belongs to a family of *t*-distributions, where each member of the family has a different number of degrees of freedom (see page 32), as shown in Figure 4.7. Analogous to a z-score, we have the variable

$$t = (\overline{y} - \mu)/\hat{\sigma}_{\overline{y}}, \tag{4.19}$$

where $\hat{\sigma}_{\overline{y}} = \hat{\sigma}_y/\sqrt{n}$ and $\hat{\sigma}_y$ is calculated from Equation 4.10. Just as with a Gaussian distribution, for a t-distribution, 95% of the area under the curve is between $t = \pm t(0.05)$ standard deviations of the mean, where $t(0.05)$ denotes the critical value of t for a statistical significance level of $p = 0.05$. In other words, (by analogy with Equation 4.15) there is a 95% chance that

$$-t(0.05) \leq \frac{\overline{y} - \mu}{\hat{\sigma}_{\overline{y}}} \leq +t(0.05). \tag{4.20}$$

However, the estimated population standard error $\hat{\sigma}_{\overline{y}}$, like the variance discussed on page 32, is based on n differences from a fixed sample mean and is therefore associated with a particular number of degrees of freedom, $n - 1$. This means that the critical value $t(0.05)$ also depends on the degrees of freedom, so we write it as $t(0.05, \nu)$. Accordingly, Equation 4.20 becomes

$$-t(0.05, \nu) \leq \frac{\overline{y} - \mu}{\hat{\sigma}_{\overline{y}}} \leq +t(0.05, \nu), \tag{4.21}$$

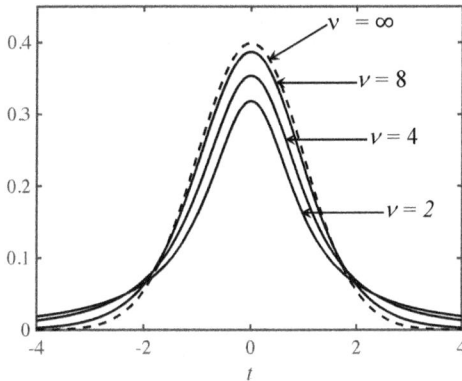

Figure 4.7: The t-distribution curves with degrees of freedom $\nu = 2$, 4 and 8. For comparison, the dashed curve is a Gaussian distribution with standard deviation $\sigma = 1$, which is close to the t-distribution with $\nu = 8$.

and (analogously to Equation 4.17) we have

$$\bar{y} - t(0.05, \nu)\,\hat{\sigma}_{\bar{y}} \quad \leq \quad \mu \quad \leq \quad \bar{y} + t(0.05, \nu)\,\hat{\sigma}_{\bar{y}}. \tag{4.22}$$

This means there is a 95% chance that the population mean μ lies between $\pm t(0.05, \nu)$ standard errors of the sample mean $\bar{y}$.

By analogy with Equation 4.18, there is a 95% chance that the population mean μ lies in the confidence interval between the confidence limits $\bar{y} - t(0.05, \nu)\hat{\sigma}_{\bar{y}}$ and $\bar{y} + t(0.05, \nu)\hat{\sigma}_{\bar{y}}$, which is written as

$$\mathrm{CI}(\mu) \quad = \quad \bar{y} \pm t(0.05, \nu)\,\hat{\sigma}_{\bar{y}}. \tag{4.23}$$

The t-distribution becomes indistinguishable from a Gaussian distribution as n (and hence ν) approaches infinity, which implies that the critical value $t(0.05, \nu)$ approaches 1.96 as n approaches infinity.

p-Values for the Mean

To find the p-value associated with the mean $\bar{y}$, look up the critical value $t(0.05, \nu)$ corresponding to $\nu = n - 1$ degrees of freedom and a significance level of $p = 0.05$ (see Table 4.1). If the absolute value of t from Equation 4.19 is larger than $t(0.05, \nu)$ (i.e. $|t| > t(0.05, \nu)$) then the p-value is $p(t, \nu) < 0.05$.

Degrees of freedom, ν	t at $p = 0.05$	t at $p = 0.01$
2	4.303	6.965
4	2.776	3.747
8	2.306	2.896
11	2.201	2.718
12	2.179	2.681
20	2.086	2.528
∞	1.960	2.326

Table 4.1: Critical values of t for a two-tailed test for different degrees of freedom ν and p-values 0.05 and 0.01, where each p-value corresponds to the total area under the two tails of the t-distribution. When $\nu = \infty$, the values of t equal the z-scores of a normalised Gaussian distribution.

Significance Versus Importance

Clearly, the lower the p-value, the more statistical significance is associated with the data. However, statistical significance does not necessarily mean anything important. For example, consider a parent population of temperature readings for which the mean is $\mu = 0.001$ degrees. If we take a sufficiently large sample of size n from this population then the difference between the sample mean $\bar{y}$ and zero will be found to be highly significant (e.g. $p < 0.00001$). Briefly, this is because the value of t in a t-test depends on the standard error of the sample mean, and this shrinks in proportion to $1/\sqrt{n}$ (Equation 4.11). Consequently, even though μ is tiny, if the sample size n is sufficiently large then t will be large (Equation 4.19), so the sample mean $\bar{y}$ will be highly significant, but it is also unimportant in this example.

This effect becomes more compelling when a t-test is used to assess the difference between two means. For example, suppose there is a difference of 1 millisecond between the mean reaction times of men and women. This would be highly significant given sufficiently large sample sizes n (e.g. if everyone on Earth was tested), but it is also unimportant.

4.9. Numerical Example

Statistical Significance of the Mean. The mean of the $n = 13$ values of y in Table 1.1 is $\bar{y} = 5.135$ feet. The variance of these values is

$$\text{var}(y) \quad = \quad \frac{1}{n}\sum_{i=1}^{n}(y_i - \bar{y})^2 \tag{4.24}$$

$$= \quad 14.240/13 = 1.095 \text{ feet}^2. \tag{4.25}$$

In contrast, the unbiased estimate of the parent population variance (Equation 4.10) involves division by the number of degrees of freedom, $\nu = n - 1 = 12$:

$$\hat{\sigma}_y^2 \quad = \quad \frac{1}{n-1}\sum_{i=1}^{n}(y_i - \bar{y})^2 \tag{4.26}$$

$$= \quad 14.240/12 = 1.187 \text{ feet}^2. \tag{4.27}$$

Therefore the estimated population standard deviation is $\hat{\sigma}_y = \sqrt{1.187} = 1.089$ feet. From Equation 4.11, the unbiased estimate of the standard error (i.e. the estimated standard deviation of the sample means) is

$$\hat{\sigma}_{\overline{y}} \;=\; \hat{\sigma}_y/\sqrt{n} = \frac{1.089}{\sqrt{13}} = 0.302 \text{ feet.} \tag{4.28}$$

If we test the null hypothesis that the data were obtained from a population with mean $\mu = 0$ then the value of t is (Equation 4.19)

$$t_{\overline{y}} \;=\; (\overline{y} - \mu)/\hat{\sigma}_{\overline{y}} = \frac{5.135 - 0}{0.302} = 16.997. \tag{4.29}$$

For $\nu = n - 1 = 13 - 1 = 12$, the critical value of t at significance level $p = 0.05$ for a one-tailed test is

$$t(0.05, 12) \;=\; 1.782, \tag{4.30}$$

where we can justify using a one-tailed test because height cannot be negative. Because $t_{\overline{y}} = 16.997$ is larger than the critical value $t(0.05, 12)$, the probability that the observed values of y could have occurred by chance given that the population mean is $\mu = 0$ is less than 5%. In fact, the exact p-value is $p = 4.617 \times 10^{-10}$; few data sets are large enough to justify this level of precision (especially a data set with only $n = 13$), so we would usually report this simply as $p < 0.01$.

Confidence Intervals of the Mean. From Equation 4.23, there is a 95% chance that the population mean μ lies in the confidence interval

$$\text{CI}(\mu) \;=\; \overline{y} \pm t(0.05, 12)\,\hat{\sigma}_{\overline{y}} \tag{4.31}$$
$$=\; 5.135 \pm (1.782 \times 0.302) \tag{4.32}$$
$$=\; 5.135 \pm 0.538. \tag{4.33}$$

Thus, there is a 95% chance that the population mean μ is between $5.135 - 0.538 = 4.597$ and $5.135 + 0.538 = 5.673$.

Reference.

Walker HM (1940). *Degrees of Freedom*. Journal of Educational Psychology, 31(4), 253.

Chapter 5

Statistical Significance: Regression

5.1. Introduction

The quality of the fit between a line and the data may look impressive, or it may look downright shoddy. In either case, how are we to assess the statistical significance of the best fitting line? In fact, this question involves two distinct subsidiary questions.

1. What is the statistical significance associated with the overall fit of the line to the data?

2. What is the statistical significance associated with each individual parameter, such as the slope and the intercept?

For reasons that will become apparent, we address these questions in reverse order. Most books on regression analysis make use of three statistical tests: the t-test, the F-test and the chi-squared test. As we shall see, the chi-squared test is not needed. And, because the t-test is a special case of the F-test, regression analysis really requires only one statistical test, the F-test. However, when assessing the statistical significance associated with individual parameters, it is convenient to use the t-test.

5.2. Statistical Significance

As stated in the previous chapter, the slope of the best fitting line is a weighted mean. Because a conventional mean is a special case of a weighted mean in which every weight equals 1, we can adapt the standard methods from the previous chapter for finding the statistical significance of means to find the statistical significance of the slope.

Slope is a Weighted Mean

We can see that the slope b_1 is a weighted mean of y values from Equation 3.26 (repeated here),

$$b_1 = \frac{\sum_{i=1}^{n} x_i \, y_i}{\sum_{i=1}^{n} x_i^2},$$ (5.1)

which can be re-written as

$$b_1 = \sum_{i=1}^{n} w_i \, y_i,$$ (5.2)

where the ith weight is

$$w_i = \frac{x_i}{\sum_{i=1}^{n} x_i^2}.$$ (5.3)

5.3. Statistical Significance: Slope

Now we return to the second question posed at the beginning of this chapter: what is the statistical significance associated with each individual parameter, such as the slope and the intercept? Specifically, how can we assess the statistical significance of the slope of the best fitting line?

Basically, the answer comes down to whether it is statistically plausible that the data could have arisen by chance alone. In the context of linear regression, this can be rephrased as the following question: what is the probability that observed data with a best fitting line of slope b_1 are just noisy versions of values that lie on a horizontal line? In this case the null hypothesis is that the relationship between the variables is a horizontal line, i.e. there is no dependence between x and y.

One way to answer this question is to estimate how different the best fitting line with a slope of b_1 is from a line with slope zero. Let us denote the null hypothesis slope by b_1', so that $b_1' = 0$. Suppose the data in Table 1.1 represent a sample of n values

$$y_i = b_1 x_i + b_0 + \eta_i,$$ (5.4)

where b_1 is the slope of the linear relationship between y and x values, b_0 is the intercept, and η_i is random noise. The presence of this noise means that any estimate of b_1 based on values of y also contains noise.

Suppose also that each sample is the result of a single run of an experiment in which we measure the values y_i corresponding to a fixed set of values $\{x_1, \ldots, x_n\}$. If we run the experiment N times then we get N different data sets where the underlying relationship between x and y remains constant, but the noise in the ith observed value y_i is different in each run of the experiment. Using these N data sets, we can obtain a set of N different estimates of the slope,

$$\{b_{11}, b_{12}, \ldots, b_{1N}\}. \tag{5.5}$$

If we plot a histogram of these N estimates then we would find that they approximate a Gaussian distribution (because each estimated value b_{1j} is a mean, which therefore obeys the central limit theorem; see page 30). Crucially, when measured in units of standard deviations, the distance between the slope b_1 estimated from a data set and the slope b_1' is

$$t_{b1} = \frac{b_1 - b_1'}{\hat{\sigma}_{b1}}, \tag{5.6}$$

where we have defined $b_1' = 0$, so that

$$t_{b1} = \frac{b_1}{\hat{\sigma}_{b1}}. \tag{5.7}$$

Finding the Estimate $\hat{\sigma}_{b1}$ of σ_{b1}. The central limit theorem (Section 4.2) implies that if each member of a set of variables $\{y_1, y_2, \ldots, y_n\}$ has a Gaussian distribution then any linear combination of those variables also has a Gaussian distribution. From Equation 5.2 we know that the slope parameter b_1 is a linear combination of y values, and we also know that each of these values has a Gaussian distribution; therefore, b_1 has a Gaussian distribution. We can obtain the variance σ_{b1}^2 of this distribution as follows. From Equation 5.2, we have

$$\sigma_{b1}^2 = \text{var}\left(\sum_{i=1}^{n} w_i y_i\right). \tag{5.8}$$

If the values of y_i are uncorrelated with each other then

$$\text{var}\left(\sum_{i=1}^{n} w_i y_i\right) = \sum_{i=1}^{n} \text{var}(w_i y_i). \tag{5.9}$$

Also, $\text{var}(w_i y_i) = w_i^2 \, \text{var}(y_i)$ (see page 17), so

$$\sigma_{b1}^2 = \sum_{i=1}^{n} w_i^2 \, \text{var}(y_i). \tag{5.10}$$

Recall that we are assuming that the variances of all the data points are the same, $\text{var}(y_1) = \text{var}(y_2) = \cdots = \text{var}(y_n)$, with all being equal to the variance σ_η^2 of the noise η_i in y_i; therefore

$$\sigma_{b1}^2 = \sigma_\eta^2 \sum_{i=1}^{n} w_i^2. \tag{5.11}$$

Using Equation 5.3, we find that $\sum_{i=1}^{n} w_i^2 = 1/\sum_{i=1}^{n}(x_i - \bar{x})^2$, so

$$\sigma_{b1}^2 = \frac{\sigma_\eta^2}{\sum_{i=1}^{n}(x_i - \bar{x})^2}. \tag{5.12}$$

Of course, we do not know the value of σ_η^2, but an unbiased estimate is

$$\hat{\sigma}_\eta^2 = \sum_{i=1}^{n} \frac{(y_i - \hat{y}_i)^2}{n-2} \tag{5.13}$$

$$= \frac{E}{n-2}, \tag{5.14}$$

where $n-2$ is the number of degrees of freedom (see Section 4.3). When assessing the number of degrees of freedom, we start with n, and then we lose one degree of freedom per parameter in the regression model. For simple regression there are $p = 2$ parameters, the slope and intercept; so the number of degrees of freedom is $\nu = n - p = n - 2$. Substituting Equation 5.13 into Equation 5.12 and taking square roots yields the unbiased estimated standard deviation of the slope b_1,

$$\hat{\sigma}_{b1} = \frac{\left[\frac{1}{n-2}\sum_{i=1}^{n}(y_i - \hat{y}_i)^2\right]^{1/2}}{\left[\sum_{i=1}^{n}(x_i - \bar{x})^2\right]^{1/2}}. \tag{5.15}$$

This can also be derived from a vector–matrix formulation (Section 7.3), which leads to the general solution in Equation 7.34.

For purely explanatory purposes, this can be simplified as follows. If n is large then $1/(n-2) \approx 1/n$, so the numerator is approximately equal to s_η, the standard deviation of the $\eta_i = y_i - \hat{y}_i$ values. Similarly, (from Equation 3.7) we recognise the denominator as $\sqrt{n \operatorname{var}(x)} = s_x\sqrt{n}$. Thus, for large n Equation 5.15 can be written as

$$\hat{\sigma}_{b1} \approx \frac{s_\eta}{s_x\sqrt{n}}. \tag{5.16}$$

Note that the standard deviation of the estimated slope is inversely proportional to the square root of the number n of data points in each sample, so it has the same form as in Figure 4.3. By analogy with Equation 4.11, the unbiased estimate of the standard deviation in the mean value of the noise is

$$\hat{\sigma}_{\bar{\eta}} = \frac{\hat{\sigma}_\eta}{\sqrt{n}}, \tag{5.17}$$

where $\hat{\sigma}_\eta \approx s_\eta$ and $\hat{\sigma}_{\bar{\eta}} \approx s_{\bar{\eta}}$ for large n, so that

$$s_{\bar{\eta}} \approx \frac{s_\eta}{\sqrt{n}}. \tag{5.18}$$

Substituting this into Equation 5.16 yields

$$\hat{\sigma}_{b1} \approx \frac{s_{\bar{\eta}}}{s_x}. \tag{5.19}$$

As s_x is constant because the x_i values are fixed, the standard deviation of the estimated slope is proportional to the standard deviation in the mean value of the noise. This makes intuitive sense, because one would expect the standard deviation of the slope of the best fitting line to depend on the amount of noise in the observed values of y.

It can be shown that Equation 5.7 can be evaluated in terms of the correlation r between x and y:

$$t_{b1} = \frac{r(n-2)^{1/2}}{(1-r^2)^{1/2}}, \tag{5.20}$$

where r is obtained from Equation 3.27 (repeated here),

$$r^2 = \frac{\text{var}(\hat{y})}{\text{var}(y)}. \tag{5.21}$$

Confidence Interval: Slope

By analogy with Equation 4.23, if b_1 is the slope of the best fitting line then there is a 95% chance that the population mean μ_{b1} lies in the confidence interval between $b_1 - t(0.05, \nu)\hat{\sigma}_{b1}$ and $b_1 + t(0.05, \nu)\hat{\sigma}_{b1}$, where $\nu = n - 2$; that is,

$$\text{CI}(\mu_{b1}) = b_1 \pm t(0.05, \nu)\,\hat{\sigma}_{b1}. \tag{5.22}$$

p-Values: Slope

The data contain noise, so it is possible that they actually have a slope of zero, even though the best fitting line has a nonzero slope b_1. The p-value is the probability that the slope b_1 of the best fitting line is due to noise in the data. More precisely, the p-value is the probability that the slope of the best fitting line is equal to or *more extreme than* the b_1 observed, given that the true slope is zero.

To find the p-value associated with the slope b_1, look up the critical value $t(0.05, \nu)$ that corresponds to $\nu = n - 2$ degrees of freedom and a significance value of $p = 0.05$ (see Table 4.1). If the absolute value of t_{b1} from Equation 5.7 is larger than $t(0.05, \nu)$ (i.e. $|t_{b1}| > t(0.05, \nu)$) then the p-value is $p(t_{b1}, \nu) < 0.05$. Notice that, in principle, the slope could have been negative or positive, so we use a two-tailed test.

> **Key point**: The data contain noise, so it is possible that they actually have a slope of zero, even though the best fitting line has nonzero slope. The p-value is the probability that the slope of the best fitting line is due to noise in the data.

5.4. Statistical Significance: Intercept

The difference between the best fitting value of b_0 and a hypothetical value b_0', in units of standard deviations, is

$$t_{b0} \quad = \quad \frac{b_0 - b_0'}{\hat{\sigma}_{b0}}, \tag{5.23}$$

where $\hat{\sigma}_{b0}$ is the unbiased estimate of the standard deviation σ_{b0} of b_0. We state without proof that (also see Section 7.5)

$$\hat{\sigma}_{b0} \quad = \quad \hat{\sigma}_\eta \times \left[\frac{1}{n} + \frac{\bar{x}^2}{\sum_{i=1}^{n}(x_i - \bar{x})^2} \right]^{1/2}, \tag{5.24}$$

where $\hat{\sigma}_\eta$ is defined by Equation 5.13. As with the slope, the number of degrees of freedom is $\nu = n - 2$. For example, to test the null hypothesis that the intercept is $b_0' = 0$, we use $t_{b0} = b_0/\hat{\sigma}_{b0}$ with $\nu = n - 2$ degrees of freedom.

Confidence Interval: Intercept

By analogy with Equation 5.22, if b_0 is the intercept of the best fitting line then there is a 95% chance that the population mean μ_{b0} lies in the confidence interval between the confidence limits $b_0 - t(0.05, \nu)\hat{\sigma}_{b0}$ and $b_0 + t(0.05, \nu)\hat{\sigma}_{b0}$, where $\nu = n - 2$, which is written as

$$\mathrm{CI}(\mu_{b0}) \quad = \quad b_0 \pm t(0.05, \nu)\,\hat{\sigma}_{b0}. \tag{5.25}$$

5.5. Significance Versus Importance

The lower the p-value, the more statistical significance is associated with the data. However, as discussed in Section 4.8, statistical significance is not necessarily associated with anything important. For example, suppose that the slope of the best fitting line $y = b_1 x + b_0$ is very small (e.g. $b_1 = 0.001$). The fact that $\hat{\sigma}_{b1} \propto 1/\sqrt{n}$ (Equation 5.16) implies that for a sufficiently large sample size n, the value of $t_{b1} = b_1/\hat{\sigma}_{b1}$ (Equation 5.7) will be large, so the slope b_1 will be found to be highly significant (e.g. $p < 0.00001$). Even so, because the slope is so small, it is probably unimportant.

In contrast, the informal notion of importance is related to the proportion of variance in y that is explained by x. Despite a significant (i.e. small) p-value, if x accounts for only a tiny proportion of variance in y then x is unimportant. In such cases, there are probably other variables that account for a large proportion of variance in y, which may be discovered using *multivariate regression* (see Chapter 7).

Key point: That the slope b_1 of the best fitting line $y = b_1x + b_0$ is statistically significant does not imply that x is an important factor in accounting for y.

5.6. Assessing the Overall Fit

The F-test. In the case of the simple linear regression considered here, the t-statistic provides all the information necessary to evaluate the fit. However, when we come to consider more general models (e.g. weighted linear regression in Chapter 7), we will find that the t-test is really a special case of a more general test, called the F-test. Thus, this section paves the way for later chapters and may be skipped on first reading.

Roughly speaking, the F-test relies on the ratio F' of two proportions, [the proportion of variance in y explained by the model] to [the proportion of variance in y not explained by the model]:

$$F' = \frac{\text{proportion of explained variance}}{\text{proportion of unexplained variance}}. \tag{5.26}$$

Clearly, larger values of F' imply a better fit of the model to the data. From Equation 5.21, the proportion of variance in y explained by the regression model is r^2, so the proportion of unexplained variance is $1 - r^2$; then Equation 5.26 becomes

$$F' = \frac{r^2}{1 - r^2}. \tag{5.27}$$

However, we need to take account of the degrees of freedom associated with the explained and unexplained variances. Accordingly, the F-

ratio is

$$F(p-1, n-p) \;=\; \frac{r^2/(p-1)}{(1-r^2)/(n-p)}, \tag{5.28}$$

where p is the number of parameters, which consist of k slopes (for k regressors, or independent variables) plus an intercept, so $p = k + 1$. This can also be expressed in terms of the the sums of squares defined in Section 3.4:

$$F(p-1, n-p) \;=\; \frac{SS_{\mathrm{Exp}}/(p-1)}{SS_{\mathrm{Noise}}/(n-p)}. \tag{5.29}$$

A table of F values lists of two different degrees of freedom, the *numerator degrees of freedom* $\nu_1 = p - 1$ and the *denominator degrees of freedom* $\nu_2 = n - p$. These names matter when using an F-table to look up a p-value. In general, the expected or mean value of F is 1. From Equation 5.28, large values of r^2 (which indicate strong correlation) correspond to large values of F.

If the value of F from Equation 5.28 is larger than the value $F(p - 1, n - p)$ listed in a look-up table (at a specified p-value, such as 0.05) then the probability that the measured correlation could have occurred by chance is less than 0.05. Modern software packages usually report the exact p-value implied by a particular value of F. In other words, if $F \gg 1$ then it is improbable that there is no underlying correlation between x and y. Precisely how improbable is given by the p-value calculated using an F-test.

The F-Test and the t-Test. As mentioned earlier, the t-test is really a special case of the F-test. With one independent variable, the number of parameters is $p = 2$, so Equation 5.28 becomes

$$F(1, n-2) \;=\; \frac{r^2}{(1-r^2)/(n-2)}. \tag{5.30}$$

If we rearrange Equation 5.20 to

$$t_{b1}(n-2) \;=\; \left(\frac{r^2}{(1-r^2)/(n-2)} \right)^{1/2} \tag{5.31}$$

then we can see that $F(1, n-2) = [t_{b1}(n-2)]^2$. In general, the value of F with 1 and $\nu = n - p$ degrees of freedom equals the square of t_{b1} with ν degrees of freedom.

So far we have been considering simple regression with only one regressor, and in this case the t-test and the F-test give exactly the same result. If we consider more than one regressor, as in multivariate regression (Chapter 7), then the t-test and the F-test are no longer equivalent and the F-test must be used.

Not Using the χ^2-Test. It is common to assess the overall fit between the data and the best fitting line using either the F-test or the χ^2-test (chi-squared test). In fact, these tests are equivalent as $n \to \infty$. However, we can usually disregard the χ^2-test, because it is the least accurate of these tests when n is not very large, which is often the case in practice.

5.7. Numerical Example

Statistical Significance of the Slope. From Equation 5.15, the unbiased estimate of the standard deviation σ_{b1} of b_1 values is

$$\hat{\sigma}_{b1} = \frac{\left[\frac{1}{n-2}\sum_{i=1}^n (y_i - \hat{y}_i)^2\right]^{1/2}}{\left[\sum_{i=1}^n (x_i - \bar{x})^2\right]^{1/2}}, \tag{5.32}$$

$$= 0.831/3.373 = 0.246. \tag{5.33}$$

To test the idea that the true slope is $b_1' = 0$ we calculate (using Equation 5.7)

$$t_{b1} = b_1/\hat{\sigma}_{b1} \tag{5.34}$$

$$= 0.764/0.246 = 3.101. \tag{5.35}$$

In a look-up table (e.g. Table 4.1), we locate the row for $\nu = n - p = 13 - 2 = 11$ degrees of freedom and find that if $t = 2.201$ then the p-value is $p = 0.05$, which is written as

$$t(0.05, 11) = 2.201. \tag{5.36}$$

Because our value of t_{b1} is larger than 2.201, its associated p-value is less than 0.05. By convention, this is reported as

$$p(3.101, 11) \quad < \quad 0.05, \tag{5.37}$$

which is read as 'the t value of 3.101 with 11 degrees of freedom implies a p-value of less than 0.05'.

In other words, if the true slope is zero then the probability of observing data with a best fitting slope of $|b_1| \geq 0.764$ is less than 5%. To put this yet another way, if we were to rerun the experiment $N = 100$ times then we expect to obtain a best fitting line whose slope has magnitude greater than or equal to 0.764 about five times.

In fact, using most modern computer software, we can do better than just state that p-value is less than some critical level. For example, given $t_{b1} = 3.101$ with $\nu = 11$, a software package output yields

$$p \quad = \quad 0.0101. \tag{5.38}$$

In words, if the true slope is zero then the probability of obtaining by chance data with a best fitting slope of $|b_1| \geq 0.764$ is $p = 0.0101$.

Confidence Interval of the Slope. From Equation 5.22, if the slope of the best fitting line is b_1, there is a 95% chance that the population mean μ_{b1} lies in the confidence interval between the confidence limits $b_1 - t(0.05, \nu)\hat{\sigma}_{b1}$ and $b_1 + t(0.05, \nu)\hat{\sigma}_{b1}$, where $\nu = n - 2 = 11$:

$$\begin{aligned} \mathrm{CI}(\mu_{b1}) &= b_1 \pm t(0.05, \nu)\hat{\sigma}_{b1} & (5.39) \\ &= 0.764 \pm (2.201 \times 0.246) & (5.40) \\ &= 0.764 \pm 0.542. & (5.41) \end{aligned}$$

Statistical Significance of the Intercept. Using Equation 5.24,

$$\begin{aligned} \hat{\sigma}_{b0} &= \left[\frac{1}{n-2} \sum_{i=1}^{n} (y_i - \hat{y}_i)^2 \right]^{1/2} \times \left[\frac{1}{n} + \frac{\overline{x}^2}{\sum_{i=1}^{n}(x_i - \overline{x})^2} \right]^{1/2} & (5.42) \\ &= 0.831 \times 0.791 & (5.43) \\ &= 0.658. & (5.44) \end{aligned}$$

	parameter value	standard error $\hat{\sigma}$	t	ν	p
slope b_1	0.764	0.246	3.101	11	0.0101
intercept b_0	3.22	0.658	4.903	11	< 0.01

r	r^2	F	ν_{Num}	ν_{Den}	p
0.683	0.466	9.617	1	11	0.0101

Table 5.1: Simple regression analysis of data in Table 1.1.

Given that the best fitting intercept is $b_0 = 3.22$, we have

$$t_{b0} \quad = \quad b_0/\hat{\sigma}_{b0} \quad = \quad 3.22/0.658 \quad = \quad 4.903. \tag{5.45}$$

From Table 4.1, the critical value of t for $\nu = 11$ degrees of freedom at significance level 0.01 is

$$t(0.01, 11) \quad = \quad 2.718. \tag{5.46}$$

Because the value of t_{b0} is larger than 2.718, its associated p-value is less than 0.01. By convention, this is reported as

$$p(4.903, 11) \quad < \quad 0.01, \tag{5.47}$$

which is read as 'the t value of 4.903 with $\nu = 11$ degrees of freedom implies a p-value of less than 0.01'.

Confidence Intervals of the Intercept. From Equation 5.25, the confidence interval for the intercept with $\nu = n - 2 = 11$ is

$$\mathrm{CI}(\mu_{b0}) \quad = \quad b_0 \pm t(0.05, \nu)\hat{\sigma}_{b0} \tag{5.48}$$
$$= \quad 3.22 \pm (2.201 \times 0.658) \tag{5.49}$$
$$= \quad 3.22 \pm 1.448. \tag{5.50}$$

Assessing the Overall Model Fit

The overall fit is represented by the correlation coefficient, and the statistical significance of the correlation coefficient is assessed using the

F-statistic in Equation 5.28 (repeated here),

$$F(p-1, n-p) \;=\; \frac{r^2/(p-1)}{(1-r^2)/(n-p)}, \qquad (5.51)$$

where r^2 is equal to the coefficient of determination. From Equation 3.27, the coefficient of determination is

$$r^2 \;=\; \frac{\text{var}(\hat{y})}{\text{var}(y)} \;=\; \frac{0.511}{1.095} \;=\; 0.466. \qquad (5.52)$$

Substituting $r^2 = 0.466$, $p-1 = 2-1 = 1$ and $n-p = 13-2 = 11$ into Equation 5.51, we get

$$F(1, 11) \;=\; \frac{0.466/1}{(1-0.466)/11} = 9.617. \qquad (5.53)$$

Using the numerator degrees of freedom ($p-1 = 1$) and the denominator degrees of freedom ($n-p = 11$), we find that the p-value is

$$p \;=\; 0.0101. \qquad (5.54)$$

This agrees with the p-value from the t-test for the slope in Equation 5.38, as it should do. This is because (in the case of simple regression) the overall model fit is determined by a single quantity r^2, which is determined by the slope b_1.

Chapter 6

Maximum Likelihood Estimation

6.1. Introduction

In this chapter we show how the least squares procedure described in previous chapters is equivalent to *maximum likelihood estimation* (MLE). In essence, the equation of a straight line is a *model* (with parameters b_1 and b_0) of how one variable y (e.g. height) changes with another variable x (e.g. salary). If we model a set of n pairs of data values x_i and y_i with a line of slope b_1 and intercept b_0, the model's estimate of the value of y at x_i is

$$\hat{y}_i \quad = \quad b_1 x_i + b_0, \tag{6.1}$$

(this is the same as Equation 1.1). As shown in Figure 6.1 (see also Figure 1.3), the vertical distance between the measured value y_i and the value $\hat{y}_i$ predicted by the 'straight line' model is represented as

$$\eta_i \quad = \quad y_i - \hat{y}_i, \tag{6.2}$$

where η_i is the amount of error or *noise* in the value y_i. If we assume that η has a Gaussian or *normal distribution* with mean zero and standard deviation σ_i then the probability of a value η_i occurring is

$$p(\eta_i) \quad = \quad k_i \, e^{-\eta_i^2/(2\sigma_i^2)}, \tag{6.3}$$

where $k_i = 1/\sqrt{2\pi\sigma_i^2}$ is a normalising constant which ensures that the area under the Gaussian distribution curve sums to 1. More accurately,

$p(\eta_i)$ is a *probability density*, but we need not worry about such subtle distinctions here. The shape of a Gaussian distribution with $\sigma = 1$ is shown in Figure 4.2 (p30).

If we substitute Equations 6.1 and 6.2 into Equation 6.3, we obtain

$$p(\eta_i) \quad = \quad k_i \, e^{-[y_i - (b_1 x_i + b_0)]^2/(2\sigma_i^2)}. \qquad (6.4)$$

But because the values of b_1 and b_0 are determined by the model, while the values of x_i are fixed, the probability $p(\eta_i)$ is also a function of y_i, which represents the probability of observing y_i, so we can replace $p(\eta_i)$ with $p(y_i)$ and write

$$p(y_i) \quad = \quad k_i \, e^{-[y_i - (b_1 x_i + b_0)]^2/(2\sigma_i^2)}. \qquad (6.5)$$

In words, the probability of observing a value y_i that contains noise η_i varies as a Gaussian function of η_i, where this Gaussian function has mean zero and standard deviation σ_i. A short way of writing that a variable η has a Gaussian or normal distribution function with mean μ and standard deviation σ is $\eta \sim \mathcal{N}(\mu, \sigma^2)$, where σ^2 is the variance of η. So, given that noise is assumed to have a mean of zero, we have $\eta_i \sim \mathcal{N}(0, \sigma_i^2)$. Notice that each value η_i can have its own unique variance σ_i^2, which indicates the reliability of the corresponding value y_i.

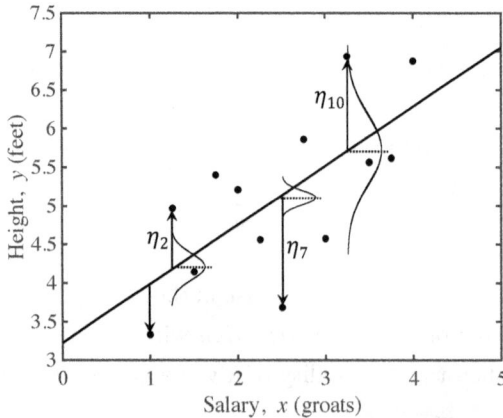

Figure 6.1: The vertical distance between each measured value y_i and the value $\hat{y}_i$ predicted by the straight line model is η_i, shown for four data points here. All measured values contain noise, and the noise of each data point is assumed to have a Gaussian distribution; each distribution has a different standard deviation, indicated by the width of the vertical Gaussian curves.

6.2. The Likelihood Function

Because the probability of observing y_i depends on the parameters b_1 and b_0, the distribution of y_i values is a *conditional probability distribution*. The conditional nature of this distribution is made explicit by using the vertical bar notation, so Equation 6.5 becomes

$$p(y_i|b_1, b_0) \quad = \quad k_i \, e^{-[y_i - (b_1 x_i + b_0)]^2/(2\sigma_i^2)}. \tag{6.6}$$

In words, $p(y_i|b_1, b_0)$ is interpreted as 'the conditional probability that the variable y has the value y_i given the parameter values b_1 and b_0'. Because $p(y_i) = p(\eta_i)$, we could have written this as $p(\eta_i|b_1, b_0)$. The conditional probability $p(y_i|b_1, b_0)$ that the variable y equals y_i is also interpreted as the *likelihood* of the parameter values b_1 and b_0.

Combining Probabilities: Coin Flipping. We usually have more than one data point, so we need to know how to combine their probabilities. To get a feel for how to do this, consider the simple example of flipping a coin. A typical coin is equally likely to yield a head or a tail, but let's imagine that this is a biased coin for which the probability of a head is given by the parameter θ. Since the probability of a head is $p(h|\theta) = \theta$, the probability of a tail is $p(t|\theta) = 1 - \theta$; the vertical bar notation makes the dependence on θ explicit.

For example, consider a coin for which the probability of a head is $\theta = 0.9$, so that the probability of a tail is $1 - \theta = 0.1$. If $n = 3$ coin flips produce a head followed by two tails, we write this result as the sequence (h, t, t). Because none of the individual flip outcomes depends on the other two flip outcomes, all outcomes are independent of each other, so the joint probability $p(h, t, t)$ of all three outcomes is obtained by multiplying the probabilities of the individual outcomes:

$$p(h, t, t|\theta) \quad = \quad p(h|\theta) \times p(t|\theta) \times p(t|\theta) \tag{6.7}$$

$$= \quad 0.9 \times 0.1 \times 0.1 \tag{6.8}$$

$$= \quad 0.009. \tag{6.9}$$

This is interpreted as the likelihood that $\theta = 0.9$.

Combining Probabilities: Regression. Analogously, when we fit a line to data, it is assumed that the noise η_i in each observed value y_i is independent of the noise in all other values of y, so the probability of obtaining a sample of n noise values $(\eta_1, \eta_2, \ldots, \eta_n)$ is

$$p(\eta_1, \eta_2, \ldots, \eta_n | b_1, b_0)$$
$$= p(\eta_1 | b_1, b_0) \times p(\eta_2 | b_1, b_0) \times \cdots \times p(\eta_n | b_1, b_0). \quad (6.10)$$

But since $p(\eta_i | b_1, b_0) = p(y_i | b_1, b_0)$, Equation 6.10 can be expressed as

$$p(y_1, y_2, \ldots, y_n | b_1, b_0) = p(y_1 | b_1, b_0) \times \cdots \times p(y_n | b_1, b_0). \quad (6.11)$$

If we write the sample of y values as a *vector* (in bold),

$$\mathbf{y} = (y_1, y_2, \ldots, y_n), \quad (6.12)$$

then Equation 6.11 becomes

$$p(\mathbf{y} | b_1, b_0) = p(y_1 | b_1, b_0) \times \cdots \times p(y_n | b_1, b_0), \quad (6.13)$$

which can be written more succinctly as the *likelihood function*

$$p(\mathbf{y} | b_1, b_0) = \prod_{i=1}^{n} p(y_i | b_1, b_0), \quad (6.14)$$

where Π, the capital Greek letter pi, is an instruction to multiply together all the terms to its right (see Appendix B). Substituting Equation 6.6 into Equation 6.14 gives

$$p(\mathbf{y} | b_1, b_0) = \prod_{i=1}^{n} k_i \, e^{-[y_i - (b_1 x_i + b_0)]^2 / (2\sigma_i^2)}, \quad (6.15)$$

where (as a reminder) $k_i = 1/(\sigma_i \sqrt{2\pi})$. In words, $p(\mathbf{y} | b_1, b_0)$ is interpreted as 'the conditional probability that the variable y adopts the set of values $\mathbf{y}$, given the parameter values b_1 and b_0'. Equation 6.15 is called the *likelihood function* because its value varies as a function of the parameters b_1 and b_0. The parameter values that make the observed data most probable are the *maximum likelihood estimate* (MLE).

The Probability of the Data? At first, this way of thinking about data seems odd. It just sounds wrong to speak of the probability of the data, which are values we have already observed, so why would we care how probable they are? In fact, we do not care about the probability of the data per se, but we do care how probable those data are in the context of the parameters we wish to estimate — that is, in the context of our regression model (Equation 6.1). Specifically, in our model, we want to find the values of the parameters b_1 and b_0 that would make the observed data most probable (i.e. the values of b_1 and b_0 that are most consistent with the data).

6.3. Likelihood and Least Squares Estimation

For reasons that will become clear, it is customary to take the logarithm of quantities like those in Equation 6.15. As a reminder, given two positive numbers m and n, $\log(m \times n) = \log m + \log n$. Accordingly, the *log likelihood* of b_1 and b_0 is

$$\log p(\mathbf{y}|b_1, b_0) = \sum_{i=1}^{n} \log \frac{1}{\sigma_i \sqrt{2\pi}} - \sum_{i=1}^{n} \frac{\left(y_i - (b_1 x_i + b_0)\right)^2}{2\,\sigma_i^2}. \qquad (6.16)$$

If we plot the value of $\log p(\mathbf{y}|b_1, b_0)$ for different putative values of b_1 and b_0 then we obtain the log likelihood function, which is qualitatively similar to the bowl-shaped function plotted in Figure 2.1.

The standard deviation σ_i of each measured value y_i effectively 'discounts' less reliable measurements, so that noisier measurements have less influence on the parameter values (b_1, b_0) of the fitted line. Notice that, except for the $1/\sigma_i$ factor, this is starting to resemble the sum of squared errors in Equation 1.25. The assumption that noise variances may not all be the same is called *heteroscedasticity*, whereas *homoscedasticity* is the assumption that all variances are the same.

If we do not know the σ_i values then we may as well assume that all of them are the same. To simplify calculations, we set $\sigma_i = 1/\sqrt{2}$, so that Equation 6.16 becomes

$$\log p(\mathbf{y}|b_1, b_0) = n\log(1/\sqrt{\pi}) - \sum_{i=1}^{n} \left(y_i - (b_1 x_i + b_0)\right)^2. \qquad (6.17)$$

Rearranging this yields

$$n \log(1/\sqrt{\pi}) - \log p(\mathbf{y}|b_1, b_0) \; = \; \sum_{i=1}^{n} \left(y_i - (b_1 x_i + b_0)\right)^2 \quad (6.18)$$

$$= \; E, \quad (6.19)$$

where the right-hand side is identical to the sum of squared errors in Equation 1.25.

Notice that the values of b_1 and b_0 that make the log likelihood $\log p(\mathbf{y}|b_1, b_0)$ (Equation 6.17) as large as possible (but always negative) also make E (Equation 6.19) as small as possible (but always positive). Thus, the values of b_1 and b_0 that *maximise* the log likelihood $\log p(\mathbf{y}|b_1, b_0)$ also *minimise* the sum of squared errors E. Therefore, if the reliabilities of all data points are the same then the maximum likelihood estimate equals the least squares estimate.

This bears repeating more fully: if all data points are equally reliable (i.e. if all σ_i values are the same) then the MLE of b_1 and b_0 is identical to the LSE. That is why the apparently arbitrary decision to estimate b_1 and b_0 by minimising the sum of squared errors turns out to be a good idea (see Section 1.3, page 5).

> **Key point**: If all data points are equally reliable (i.e. if all σ_i values are the same) then the maximum likelihood estimate (MLE) of model parameter values is identical to the least squares estimate (LSE) of those parameter values.

Chapter 7

Multivariate Regression

7.1. Introduction

Most measured quantities depend on several other quantities. For example, a person's score on a new computer game *zog* depends on a combination of the number x_1 of hours they spent playing zog and the number x_2 of years they spent playing computer games in general. For a sample of n individuals, the score of the ith individual can be modelled as

$$y_i \;=\; b_1 x_{i1} + b_2 x_{i2} + b_0 + \eta_i. \qquad (7.1)$$

The parameter b_1 specifies the extent to which the score y_i depends on the number of hours spent playing zog, b_2 specifies the extent to which the score depends on previous experience with computer games, b_0 represents the average score of naive individuals (i.e. individuals who have not spent any time playing zog and who have no experience with computer games), and η_i is the noise in the score measurements.

7.2. The Best Fitting Plane

Just as simple regression has a geometric interpretation as fitting a straight line to data, so multivariate regression can be interpreted as fitting a plane to data. In the case of two independent variables x_1 and x_2, each location (x_{i1}, x_{i2}) on the 'ground' has a corresponding point at height $\hat{y}_i$ on the best fitting plane with equation

$$\hat{y}_i \;=\; b_1 x_{i1} + b_2 x_{i2} + b_0, \tag{7.2}$$

as shown in Figure 7.1.

Note that the value of $\hat{y}$ depends on $p = 3$ parameters, comprising $k = 2$ regression coefficients

$$b_1 \;=\; \frac{\Delta \hat{y}}{\Delta x_1} \tag{7.3}$$

$$b_2 \;=\; \frac{\Delta \hat{y}}{\Delta x_2} \tag{7.4}$$

plus one intercept parameter b_0. The vertical distance between each measured data point y_i and the corresponding point $\hat{y}_i$ on the best fitting plane is considered to be measurement error or noise, $\eta_i = y_i - \hat{y}_i$.

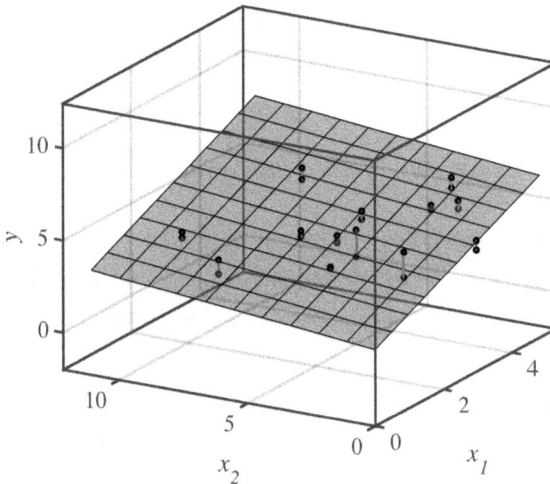

Figure 7.1: Given a data set consisting of x_1, x_2 and y values, where the variable y is thought to depend on the variables x_1 and x_2, we can fit a plane to the data, defined by $\hat{y}_i = b_1 x_{i1} + b_2 x_{i2} + b_0$, where b_1 is the gradient with respect to x_1 (i.e. the slope of the plane along the wall formed by the x_1- and y-axes), b_2 is the gradient with respect to x_2 (the slope of the plane along the wall formed by the x_2- and y-axes), and b_0 is the height of the plane at $(x_1, x_2) = (0, 0)$ on the ground. Each vertical line joins a measured data point y_i to the point $\hat{y}_i$ on the best fitting plane above the same location $(x_{i1}; x_{i2})$ on the ground. Using the data in Table 7.1, the least squares estimates are $b_1 = 0.966$, $b_2 = 0.138$ and $b_0 = 2.148$.

i	1	2	3	4	5	6	7	8	9	10	11	12	13
y_i	3.34	4.97	4.15	5.40	5.21	4.56	3.69	5.86	4.58	6.94	5.57	5.62	6.87
x_{i1}	1.00	1.25	1.50	1.75	2.00	2.25	2.50	2.75	3.00	3.25	3.50	3.75	4.00
x_{i2}	7.47	9.24	3.78	1.23	5.57	4.48	4.05	4.19	0.05	7.20	2.48	1.73	2.37

Table 7.1: Values of two independent variables (regressors) x_1 and x_2 and a dependent variable y.

And, just as simple regression finds the best fitting line by minimising the sum of squared vertical distances between the line and the y data, so multivariate regression finds the best fitting *plane* by minimising the sum of squared vertical distances between the plane and the y data,

$$E = \sum_{i=1}^{n}(y_i - \hat{y}_i)^2 \qquad (7.5)$$

$$= \sum_{i=1}^{n}\left(y_i - (b_1 x_{i1} + b_2 x_{i2} + b_0)\right)^2, \qquad (7.6)$$

In summary, for $k = 2$ regressors (independent variables), the best fitting two-dimensional model is a plane embedded in three-dimensional space, as in Figure 7.1. More generally, if there are $k > 2$ regressors then the best fitting k-dimensional model is a hyper-plane embedded in $(k + 1)$-dimensional space.

> **Key point**: Just as simple regression finds the best fitting line by minimising the sum of squared vertical distances between the line and the data, so multivariate regression finds the best fitting *plane* by minimising the sum of squared vertical distances between the plane and the data.

7.3. Vector–Matrix Formulation

Expressing regression in terms of vectors provides a notational uniformity which ensures that extension beyond two regressors is reasonably self-evident (see Appendix C). The right-hand side of Equation 7.2 can be expressed as the *inner product* of the *column vector*

$$\mathbf{b} = \begin{pmatrix} b_1 \\ b_2 \\ b_0 \end{pmatrix} \qquad (7.7)$$

and the *row vector*

$$\mathbf{x}_i = (x_{i1}, x_{i2}, 1), \qquad (7.8)$$

where the first two elements of $\mathbf{x}_i$ define a single location on the ground plane. So Equation 7.2 can be written in vector–matrix form as

$$\hat{y}_i = (x_{i1}, x_{i2}, 1) \begin{pmatrix} b_1 \\ b_2 \\ b_0 \end{pmatrix} \qquad (7.9)$$

or, more succinctly,

$$\hat{y}_i = \mathbf{x}_i \mathbf{b}. \qquad (7.10)$$

When considered over all n data points, we have

$$\begin{pmatrix} \hat{y}_1 \\ \vdots \\ \hat{y}_n \end{pmatrix} = \begin{pmatrix} \mathbf{x}_1 \\ \vdots \\ \mathbf{x}_n \end{pmatrix} \mathbf{b}. \qquad (7.11)$$

The n values of $\hat{y}$ can be represented as a column vector

$$\hat{\mathbf{y}} = \begin{pmatrix} \hat{y}_1 \\ \vdots \\ \hat{y}_n \end{pmatrix}, \qquad (7.12)$$

and the term in brackets on the right-hand side of Equation 7.11 can be represented as the $n \times 3$ matrix

$$X \;=\; \begin{pmatrix} \mathbf{x}_1 \\ \vdots \\ \mathbf{x}_n \end{pmatrix}, \tag{7.13}$$

where each row can be expanded to yield

$$X \;=\; \begin{pmatrix} x_{11} & x_{12} & 1 \\ \vdots & \cdots & \vdots \\ x_{n1} & x_{n2} & 1 \end{pmatrix}. \tag{7.14}$$

When Equation 7.11 is written out in full,

$$\begin{pmatrix} \hat{y}_1 \\ \vdots \\ \hat{y}_n \end{pmatrix} \;=\; \begin{pmatrix} x_{11} & x_{12} & 1 \\ \vdots & \cdots & \vdots \\ x_{n1} & x_{n2} & 1 \end{pmatrix} \begin{pmatrix} b_1 \\ b_2 \\ b_0 \end{pmatrix}, \tag{7.15}$$

it becomes apparent that we have n simultaneous equations and three unknowns (b_1, b_2 and b_0), which can now be written succinctly as

$$\hat{\mathbf{y}} \;=\; X\mathbf{b}. \tag{7.16}$$

This can be used to express Equation 7.5 in vector–matrix format, as follows. If we write the n measured values of y as a column vector

$$\mathbf{y} \;=\; \begin{pmatrix} y_1 \\ \vdots \\ y_n \end{pmatrix}, \tag{7.17}$$

then Equation 7.5 can be written as

$$E \;=\; (\mathbf{y} - \hat{\mathbf{y}})^\mathsf{T}(\mathbf{y} - \hat{\mathbf{y}}), \qquad (7.18)$$

where $\mathbf{y}$ and $\hat{\mathbf{y}}$ are vectors of n elements and the *transpose operator* T converts column vectors to row vectors (and vice versa). From Equation 7.16, we have

$$E \;=\; (\mathbf{y} - X\mathbf{b})^\mathsf{T}(\mathbf{y} - X\mathbf{b}), \qquad (7.19)$$

and expanding this yields

$$E \;=\; \mathbf{y}^\mathsf{T}\mathbf{y} + \mathbf{b}^\mathsf{T}X^\mathsf{T}X\mathbf{b} - 2\mathbf{b}^\mathsf{T}X^\mathsf{T}\mathbf{y}, \qquad (7.20)$$

where the final (cross) term results from multiplying $X\mathbf{b}$ by $\mathbf{y}$ twice and using the transpose property $(X\mathbf{b})^\mathsf{T} = \mathbf{b}^\mathsf{T}X^\mathsf{T}$ (see Appendix C).

7.4. Finding the Best Fitting Plane

The usual method for finding the regression coefficients $\mathbf{b}$ consists of taking the derivative of E with respect to each regression coefficient and then using these derivatives to find the values of the regression coefficients that make the slope zero.

The intuition behind this method is that at a minimum of the function E with respect to $\mathbf{b}$, the slope must be zero. Turning this around, if we find a value of $\mathbf{b}$ which makes the slope zero then we automatically have a value for $\mathbf{b}$ that may give a minimum of E (of course, it could also correspond to a maximum, but there are ways of checking whether it is a minimum as we want).

The derivative with respect to the 3-element column vector $\mathbf{b} = (b_1, b_2, b_0)^\mathsf{T}$ is also a 3-element column vector:

$$\nabla E \;=\; \left(\frac{\partial E}{\partial b_1}, \frac{\partial E}{\partial b_2}, \frac{\partial E}{\partial b_0} \right)^\mathsf{T}, \qquad (7.21)$$

where the *nabla* symbol ∇ is standard notation for the gradient operator on a vector, i.e. $\nabla E = dE/d\mathbf{b}$. Each element of the vector ∇E is the derivative of E with respect to one regression coefficient, and at a minimum of E all these derivatives must equal zero:

$$\nabla E \;=\; (0,0,0)^{\mathsf{T}}. \tag{7.22}$$

The derivative of Equation 7.20 is

$$\nabla E \;=\; 2[(X^{\mathsf{T}}X)\mathbf{b} - X^{\mathsf{T}}\mathbf{y}]. \tag{7.23}$$

At a minimum this equals the zero vector in Equation 7.22, which yields

$$(X^{\mathsf{T}}X)\mathbf{b} \;=\; X^{\mathsf{T}}\mathbf{y}. \tag{7.24}$$

Here $X^{\mathsf{T}}X$ is a 3×3 covariance matrix, and $X^{\mathsf{T}}\mathbf{y}$ is a 3×1 column vector. As in Section 2.4, Equation 7.24 represents three simultaneous equations in three unknowns, $\mathbf{b} = (b_1, b_2, b_0)^{\mathsf{T}}$.

Multiplying both sides of Equation 7.24 by $(X^{\mathsf{T}}X)^{-1}$ yields the least squares estimate of the regression coefficients,

$$\mathbf{b} \;=\; (X^{\mathsf{T}}X)^{-1}X^{\mathsf{T}}\mathbf{y}, \tag{7.25}$$

where $(X^{\mathsf{T}}X)^{-1}$ is the 3×3 inverse of the matrix $X^{\mathsf{T}}X$.

From Equation 7.16 ($\hat{\mathbf{y}} = X\mathbf{b}$), we obtain $\hat{\mathbf{y}}$ as

$$\hat{\mathbf{y}} \;=\; X(X^{\mathsf{T}}X)^{-1}X^{\mathsf{T}}\mathbf{y}, \tag{7.26}$$

where the $n \times n$ matrix $H = X(X^{\mathsf{T}}X)^{-1}X^{\mathsf{T}}$ is called the *hat matrix*. Note that this expression for $\hat{\mathbf{y}}$ involves only the x and y values from the data.

Standardised Regression Coefficients

Suppose we find that the regression coefficient b_1 is larger than the coefficient b_2. Does that mean that the regressor x_1 is more important than the regressor x_2? The answer depends on the magnitudes of x_1 and x_2. For example, if we were to halve all values of x_1 and repeat the regression analysis, we would find that the value of b_1 doubles but the value of b_2 remains unchanged. Clearly, halving all values of the regressor x_1 did not suddenly make it more important in relation to the regressor x_2. Consequently, we cannot simply compare the raw coefficients b_1 and b_2 without taking into account the magnitudes of the corresponding regressors x_1 and x_2.

In order to compare different regression coefficients, we first need to ensure that the corresponding regressors have the same standard deviation. This is achieved by dividing each variable by its standard deviation, which defines new *standardised* or *normalised* variables

$$x_1' = x_1/s_{x1}, \qquad x_2' = x_2/s_{x2}, \qquad y' = y/s_y, \qquad (7.27)$$

where s represents standard deviation. Thus each of the standardised variables x_1', x_2' and y' has a standard deviation of 1. Now the normalised regressors x_1' and x_2' have the same intrinsic scale, so their coefficients can be compared on an equal footing.

> **Key point:** The relative importance of different regressors can only be assessed by comparing the magnitudes of their *standardised* regression coefficients.

Multicollinearity

From the equation of multivariate regression, it is apparent that the contribution of each regressor to y can be added to the contributions of all other regressors. But if two different regressors are correlated then the equation effectively 'double counts' their contributions to y. For this reason, multivariate regression is based on the assumption that the regressors are mutually uncorrelated. Multicollinearity is a measure of the extent to which regressors are correlated with each other.

7.5. Statistical Significance

Degrees of Freedom

For multivariate regression the parameters include k regressors and the intercept b_0, so the number p of parameters is $p = k+1$. When assessing the number of degrees of freedom ν, we lose one degree of freedom per regressor, and we also lose one degree of freedom for the intercept; so the number of degrees of freedom for p parameters is $\nu = n - p$.

Assessing the Overall Model Fit

The proportion of the variance in y that can be attributed to the best fitting plane is the *coefficient of determination*, which equals the square of the correlation coefficient (Equation 3.27, repeated here):

$$r^2 = \frac{\text{var}(\hat{y})}{\text{var}(y)}. \tag{7.28}$$

In the context of multivariate regression, r is the *multiple correlation coefficient*, and r^2 is a measure of how well the best fitting plane 'soaks up' the variance in y, i.e. the proportion of variance in y explained by the regression model. From Equations 3.17–3.23, we can also express this as

$$r^2 = 1 - \frac{\sum_{i=1}^{n}(y_i - \hat{y}_i)^2}{\sum_{i=1}^{n}(y_i - \overline{y})^2}. \tag{7.29}$$

Adding more regressors usually increases r^2, sometimes by pure chance. As an extreme example, suppose we added 1,000 regressors, where each regressor is a random set of n values. Almost inevitably, some of these regressors are correlated with y, and will therefore increase r^2. Accordingly, we can take account of the number of regressors by using the *adjusted r^2 statistic*

$$r_{\text{Adj}}^2 = 1 - \frac{1/(n-p)\sum_{i=1}^{n}(\hat{y}_i - y_i)^2}{1/(n-1)\sum_{i=1}^{n}(y_i - \overline{y})^2}, \tag{7.30}$$

where p is the number of parameters (k regressors plus the intercept).

The statistical significance of the coefficient of determination is assessed using the F-ratio (Equation 5.28, repeated here) with numerator degrees of freedom $p - 1$ and denominator degrees of freedom $n - p$:

$$F(p-1, n-p) = \frac{r^2/(p-1)}{(1-r^2)/(n-p)}. \tag{7.31}$$

As discussed in Section 5.6, this F-ratio implies a particular p-value. Of course, we could equivalently have used Equation 5.29 here.

Statistical Significance of Individual Parameters

To test the significance of individual parameters, we need to estimate the underlying standard deviation of each parameter. For example, the regressor parameter b_1 has an associated t-value (see Equation 5.7) of

$$t_{b1}(\nu) \quad = \quad \frac{b_1}{\hat{\sigma}_{b1}}, \tag{7.32}$$

where $\nu = n - p$ with $p = 3$ (the number of parameters). Each of the standard deviations $\hat{\sigma}_{b1}$, $\hat{\sigma}_{b2}$ and $\hat{\sigma}_{b0}$ is the square root of one diagonal element of the covariance matrix (from Equation 2.8 in Gujarati, 2019)

$$\hat{\sigma}_X^2 \quad = \quad \hat{\sigma}_\eta^2 (X^\mathsf{T} X)^{-1} \tag{7.33}$$

$$= \quad \begin{pmatrix} \hat{\sigma}_{b1}^2 & 0 & 0 \\ 0 & \hat{\sigma}_{b2}^2 & 0 \\ 0 & 0 & \hat{\sigma}_{b0} \end{pmatrix}, \tag{7.34}$$

where $\hat{\sigma}_\eta^2$ is the estimated value of the noise variance σ_η^2 (Equation 5.14),

$$\hat{\sigma}_\eta^2 \quad = \quad \frac{E}{n-p}, \tag{7.35}$$

so that Equation 7.34 can be computed as

$$\hat{\sigma}_X^2 \quad = \quad \left(\frac{E}{n-p} \right) (X^\mathsf{T} X)^{-1}. \tag{7.36}$$

7.6. How Many Regressors?

In practice, the quality of the fit is assessed using the *extra sum-of-squares* method, also known as the *partial F-test*. In essence, this consists of assessing the extra sum-of-squares SS_{Exp} (see Equation 3.19) accounted for by the regressors when the number of regressors is increased. As the number of regressors is increased, the proportion of the total sum of squared errors that is explained by the regressors increases. For example, if we consider only one regressor, i.e. $p = 2$ parameters (slope and intercept), the predicted value of y_i obtained from the *reduced model* is

$$\hat{y}_i(\mathbf{b}_{\mathrm{RED}}) \;=\; b_1 x_{i1} + b_0, \tag{7.37}$$

where the slope b_1 and intercept b_0 are represented by the vector

$$\mathbf{b}_{\mathrm{RED}} \;=\; (b_1, b_0). \tag{7.38}$$

From Equation 3.19, the sum of squared errors explained by the regressor x_{i1} is

$$SS_{\mathrm{Exp}}(\mathbf{b}_{\mathrm{RED}}) \;=\; \sum_{i=1}^{n} \big(\hat{y}_i(\mathbf{b}_{\mathrm{RED}}) - \overline{y}\big)^2, \tag{7.39}$$

where the dependence on $\mathbf{b}_{\mathrm{RED}}$ is made explicit.

If we now consider $k = 2$ regressors, x_{i1} and x_{i2}, then the predicted value of y_i from the *full model* becomes

$$\hat{y}_i(\mathbf{b}_{\mathrm{FULL}}) \;=\; b_1 x_{i1} + b_2 x_{i2} + b_0, \tag{7.40}$$

where the two slopes b_1 and b_2 and the intercept b_0 are represented as

$$\mathbf{b}_{\mathrm{FULL}} \;=\; (b_1, b_2, b_0). \tag{7.41}$$

Incidentally, the values of the coefficients b_1 and b_0 usually change when a new regressor x_{i2} is introduced. The important point is that the sum of squared errors explained by the regressors *increases* to

$$SS_{\mathrm{Exp}}(\mathbf{b}_{\mathrm{FULL}}) \;=\; \sum_{i=1}^{n} \big(\hat{y}_i(\mathbf{b}_{\mathrm{FULL}}) - \overline{y}\big)^2, \tag{7.42}$$

such that

$$SS_{\mathrm{Exp}}(\mathbf{b}_{\mathrm{FULL}}) \;\geq\; S_{\mathrm{Exp}}(\mathbf{b}_{\mathrm{RED}}). \tag{7.43}$$

The Extra Sum-of-Squares Method. Suppose we begin with the *full model* containing $p = 3$ parameters, so the sum of squared errors explained by the model is $SS_{\mathrm{Exp}}(\mathbf{b}_{\mathrm{FULL}})$ (Equation 7.42). The sum of squared errors explained by the *reduced model* with only $p = 2$ parameters is $SS_{\text{Exp}}(\mathbf{b}_{\text{RED}})$ (Equation 7.39). So upon removing one of the regressors x_2 from the full model, the *extra sum-of-squares* that had been explained by b_2 (strictly speaking, x_2) in the full model is

$$SS_{\text{Exp}}(b_2|\mathbf{b}_{\text{RED}}) = SS_{\text{Exp}}(\mathbf{b}_{\text{FULL}}) - SS_{\text{Exp}}(\mathbf{b}_{\text{RED}}). \quad (7.44)$$

To test the hypothesis that b_2 equals zero (i.e. that removing x_2 from the full model does not change the explained sum of squared errors), we calculate the F-ratio

$$F(\nu_{\text{Diff}}, n - p) = \frac{[SS_{\text{Exp}}(\mathbf{b}_{\text{FULL}}) - SS_{\text{Exp}}(\mathbf{b}_{\text{RED}})]/\nu_{\text{Diff}}}{SS_{\text{Noise}}/(n - p)}$$

$$= \frac{SS_{\text{Exp}}(b_2|\mathbf{b}_{\text{RED}})/\nu_{\text{Diff}}}{SS_{\text{Noise}}/(n - p)}, \quad (7.45)$$

where $SS_{\text{Noise}} = E$ (Equations 7.5 and 7.18), ν_{Diff} is the number of parameters in the full model minus the number of parameters in the reduced model ($\nu_{\text{Diff}} = 3 - 2 = 1$ here), and $p = k + 1$ is the number of parameters in the full model (so $p = 3$ here).

As usual, if the observed value $F(\nu_{\text{Diff}}, n - p)$ is larger than the corresponding value of F at a significance level of $p = 0.05$ then we reject the hypothesis that $b_2 = 0$, i.e. we conclude that the regressor associated with b_2 contributes significantly to the model. More generally, we may wish to test the effect of simultaneously removing several regressors from the full model, in which case $\nu_{\text{Diff}} > 1$.

7.7. Numerical Example

Finding the Best Fitting Plane. The best fitting plane for the data in Table 7.1 is shown in Figure 7.1. The least squares estimates of the regression coefficients are

$$b_1 = 0.966 \quad (\hat{\sigma}_{b1} = 0.281),$$
$$b_2 = 0.138 \quad (\hat{\sigma}_{b2} = 0.103),$$
$$b_0 = 2.148 \quad (\hat{\sigma}_{b0} = 1.022).$$

Substituting these values into Equation 7.2 gives

$$\hat{y}_i \;=\; 0.966 \times x_{i1} + 0.138 \times x_{i2} + 2.148. \qquad (7.46)$$

This seems to imply that the regressor x_1 is $0.966/0.138 = 7.00$ times more influential than x_2 on the value of y (but see below).

Standardised Regression Coefficients. Using the standardised variables $x'_{b_1} = x_1/s_{b_1}$ and $x'_{b_2} = x_2/s_{b_2}$ introduced on page 72, we obtain the parameter values $b_1 = 0.864$, $b_2 = 0.338$ and $b_0 = 2.047$, so Equation 7.2 becomes

$$\hat{y}_i \;=\; 0.864 \times x'_{i1} + 0.338 \times x'_{i1} + 2.047. \qquad (7.47)$$

This implies that the regressor x_1 is only $0.864/0.338 = 2.56$ times more influential than x_2 on the value of y.

Assessing the Overall Model Fit. Evaluating the coefficient of determination (i.e. the square of the multiple correlation coefficient, Equation 7.28), we obtain

$$r^2 \;=\; \frac{\mathrm{var}(\hat{y})}{\mathrm{var}(y)} = \frac{0.600}{1.095} = 0.548, \qquad (7.48)$$

so the multiple correlation coefficient is $r = \sqrt{0.548} = 0.740$.

The statistical significance of the multiple correlation coefficient is assessed using the F-ratio (see Section 5.6). From Equation 7.31 (repeated below), the F-ratio of the coefficient of determination with numerator degrees of freedom $p-1 = 3-1 = 2$ and denominator degrees of freedom $n - p = 13 - 3 = 10$ is

$$F(p-1, n-p) \;=\; \frac{r^2/(p-1)}{(1-r^2)/(n-p)} \qquad (7.49)$$

$$=\; \frac{0.548/2}{(1-0.548)/10} \qquad (7.50)$$

$$=\; 6.063, \qquad (7.51)$$

which corresponds to a p-value of

$$p \;=\; 0.019. \tag{7.52}$$

As this is less than 0.05, the coefficient of determination (i.e. the overall fit) is statistically significant.

Statistical Significance of the Individual Parameters. Using Equation 7.32 with $\nu = n - p = 13 - 3 = 10$ degrees of freedom, the value of t associated with the regression coefficient b_1 is

$$t_{b1} \;=\; \frac{b_1}{\hat{\sigma}_{b1}} \;=\; \frac{0.966}{0.281} \;=\; 3.433, \tag{7.53}$$

which corresponds to a p-value of $p = 0.0064$, so the slope $b_1 = 0.966$ is statistically significant.

Similarly, with $\nu = n - p = 13 - 3 = 10$ degrees of freedom, the value of t associated with the regression coefficient b_2 is

$$t_{b2} \;=\; \frac{b_2}{\hat{\sigma}_{b2}} \;=\; \frac{0.138}{0.103} \;=\; 1.344, \tag{7.54}$$

which corresponds to a p-value of $p = 0.209$, so the slope $b_2 = 0.138$ is not statistically significant.

Finally, the value of t associated with the intercept b_0 is

$$t_{b0} \;=\; \frac{b_0}{\hat{\sigma}_{b0}} \;=\; \frac{2.148}{1.022} \;=\; 2.102, \tag{7.55}$$

	parameter value	standard error $\hat{\sigma}$	t	ν	p
coefficient b_1	0.966	0.281	3.43	10	0.0064
coefficient b_2	0.138	0.103	1.34	10	0.209
intercept b_0	2.148	1.022	2.10	10	0.062
r^2	adjusted r^2	F	ν_{Num}	ν_{Den}	p
0.548	0.458	6.062	2	10	0.0189

Table 7.2: Results of multivariate regression analysis of the data in Table 7.1.

which corresponds to a p-value of $p = 0.062$, so the intercept $b_0 = 2.148$ is not statistically significant (i.e. not significantly different from zero).

Using the extra sum-of-squares method to assess if $b_2 = 0$. To assess whether the regressor x_2 contributes significantly to the model, we need to test the null hypothesis that the coefficient $b_2 = 0$. This involves calculating the two sums of squared errors in Equation 7.44:

$$SS_{\text{Exp}}(b_2|\mathbf{b}_{\text{RED}}) \quad = \quad SS_{\text{Exp}}(\mathbf{b}_{\text{FULL}}) - SS_{\text{Exp}}(\mathbf{b}_{\text{RED}}) \qquad (7.56)$$

$$= \quad 7.805 - 6.643 \qquad\qquad\qquad (7.57)$$

$$= \quad 1.162. \qquad\qquad\qquad\qquad\quad (7.58)$$

Then Equation 7.45 (repeated here),

$$F(\nu_{\text{Diff}}, n - p) \quad = \quad \frac{SS_{\text{Exp}}(b_2|\mathbf{b}_{\text{RED}})/\nu_{\text{Diff}}}{SS_{\text{Noise}}/(n - p)}, \qquad (7.59)$$

becomes

$$F(1, 10) \quad = \quad \frac{1.162/1}{6.436/10} \qquad\qquad (7.60)$$

$$= \quad 1.805, \qquad\qquad\qquad (7.61)$$

which corresponds to a p-value of 0.209 (not significant), so we cannot reject the hypothesis that $b_2 = 0$. In other words, the regressor x_2 with coefficient b_2 does not contribute significantly to the model.

Reference.

Gujarati DN (2019) *The Linear Regression Model*, Sage Publications.

Chapter 8

Weighted Linear Regression

8.1. Introduction

The method of weighted linear regression has been delayed until now because it follows naturally from the methods described in the previous two chapters. Whereas *simple linear regression* is based on the assumption that all data points are equally reliable, weighted linear regression takes account of the fact that some data points are more reliable than others, and that these should have a greater influence on the slope and intercept of the best fitting line (see 'A Line Suspended on Springs' on page 7).

8.2. Weighted Sum of Squared Errors

Consider Equation 6.16 (repeated here),

$$\log p(y|b_1, b_0) = \sum_{i=1}^{n} \log \frac{1}{\sigma_i \sqrt{2\pi}} - \frac{1}{2} \sum_{i=1}^{n} \frac{\left(y_i - (b_1 x_i + b_0)\right)^2}{\sigma_i^2}. \tag{8.1}$$

The second summation is a *weighted sum of squared errors*

$$E = \sum_{i=1}^{n} \frac{\left(y_i - (b_1 x_i + b_0)\right)^2}{\sigma_i^2} \tag{8.2}$$

$$= \left(2 \sum_{i=1}^{n} \log \frac{1}{\sigma_i \sqrt{2\pi}}\right) - 2 \log p(y|b_1, b_0). \tag{8.3}$$

Equation 8.2 can be recognised as a generalisation of the sum of squared errors defined in Equation 1.25, such that the ith squared difference is

i	1	2	3	4	5	6	7	8	9	10	11	12	13
x_i	1.00	1.25	1.50	1.75	2.00	2.25	2.50	2.75	3.00	3.25	3.50	3.75	4.00
y_i	3.34	4.97	4.15	5.40	5.21	4.56	3.69	5.86	4.58	6.94	5.57	5.62	6.87
σ_i	0.09	0.15	0.24	0.36	0.50	0.67	0.87	1.11	1.38	1.68	2.03	2.41	2.83

Table 8.1: Values of salary x_i and measured height y_i as in Table 1.1, where each value of y_i has a different standard deviation σ_i.

now 'discounted' by the noise variance σ_i^2 associated with the measured value y_i. The values of b_1 and b_0 obtained by minimising E are called the *weighted least squares estimates* (WLSE) of the slope and intercept.

As an overview of this chapter, Figure 8.1 compares the results of simple regression and weighted regression, for the data shown in Table 8.1. Whereas simple regression treats all data points as if they have the same standard deviation, weighted regression takes account of the different standard deviations of data points (also shown in Figure 6.1). The data points on the left of Table 8.1 have small standard deviations, so in weighted regression they have a relatively large effect on the best fitting line.

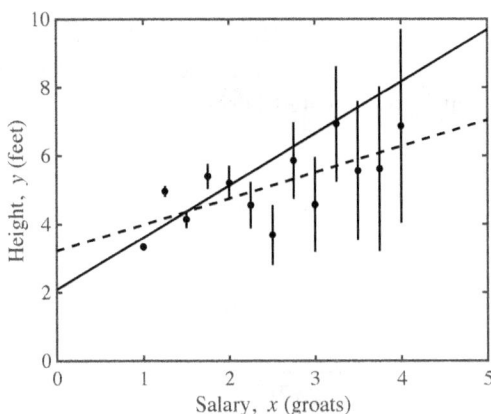

Figure 8.1: Weighted versus simple regression using the data from Table 8.1. Weighted regression yields a slope of $b_1 = 1.511$ and an intercept of $b_0 = 2.122$ (solid line); the length of each vertical line is twice the standard deviation of the respective data point. For comparison, simple regression assumes that all data points have the same standard deviation, which yields $b_1 = 0.764$ and $b_0 = 3.22$ (dashed line).

8.3. Vector–Matrix Formulation

In Section 7.3 we saw how multivariate linear regression can be expressed in terms of vectors and matrices. This approach is especially useful when dealing with data where different data points have different variances.

Equation 8.2 can be expressed in vector–matrix notation using an $n \times n$ matrix that contains n variances along its diagonal,

$$V = \begin{pmatrix} \sigma_1^2 & \cdots & 0 \\ \vdots & \ddots & \vdots \\ 0 & \cdots & \sigma_n^2 \end{pmatrix}. \tag{8.4}$$

The inverse of V is obtained by simply inverting each diagonal element,

$$W = \begin{pmatrix} 1/\sigma_1^2 & \cdots & 0 \\ \vdots & \ddots & \vdots \\ 0 & \cdots & 1/\sigma_n^2 \end{pmatrix}. \tag{8.5}$$

Making use of $\hat{\mathbf{y}}$ and $\mathbf{y}$ defined in Equations 7.12 and 7.17, we write Equation 8.2 (i.e. the unexplained sum of squares) as

$$E = (\mathbf{y} - \hat{\mathbf{y}})^\mathsf{T} W (\mathbf{y} - \hat{\mathbf{y}}). \tag{8.6}$$

From Equation 7.16, $\hat{\mathbf{y}} = X\mathbf{b}$, so Equation 8.6 becomes

$$E = (\mathbf{y} - X\mathbf{b})^\mathsf{T} W (\mathbf{y} - X\mathbf{b}). \tag{8.7}$$

The gradient of E is a two-element vector,

$$\nabla E = \left(\frac{\partial E}{\partial b_1}, \frac{\partial E}{\partial b_0} \right)^\mathsf{T}, \tag{8.8}$$

where each element is the derivative with respect to one parameter. Taking the derivative of Equation 8.7, we obtain

$$\nabla E = 2[(X^\mathsf{T} W X)\mathbf{b} - X^\mathsf{T} W \mathbf{y}]. \tag{8.9}$$

At a minimum, each element of ∇E equals zero, so

$$(X^{\mathsf{T}}WX)\mathbf{b} \;=\; X^{\mathsf{T}}W\mathbf{y}, \qquad (8.10)$$

where $X^{\mathsf{T}}WX$ is a 2×2 covariance matrix and $X^{\mathsf{T}}W\mathbf{y}$ is a 2×1 column vector. Rearranging yields the regression coefficients

$$\mathbf{b} \;=\; (X^{\mathsf{T}}WX)^{-1}X^{\mathsf{T}}W\mathbf{y}, \qquad (8.11)$$

where $\mathbf{b}$ is a column vector, $\mathbf{b} = (b_1, b_0)^{\mathsf{T}}$.

Weighted Linear Multivariate Regression. By using vector–matrix notation, *weighted linear multivariate regression* is a reasonably simple extension of the method described above. Specifically, with k regressors the matrix X would have $p = k + 1$ columns (one column per regressor plus a column of ones, Equation 7.14). The rest of the calculations above would remain the same.

8.4. Statistical Significance

Assessing the Overall Model Fit. The fit of the model to the data can be assessed using the F-ratio, which is defined as

$$F(p - 1, n - p) \;=\; \frac{r_w^2/(p - 1)}{(1 - r_w^2)/(n - p)}, \qquad (8.12)$$

where we have replaced r^2 in Equation 7.31 with r_w^2 (defined below). These are equivalent only if all values of y have the same variance, which is not the case in weighted linear regression.

The Coefficient of Determination. In the following, the development will parallel that in Chapter 3. For each data point, the total error is the difference between the observed value y_i and the mean $\bar{y}$. However, if each value of y_i has its own variance then the best estimate of the mean is no longer the conventional average of the y_i values. Instead, it should take into account the variance of each y_i, such that more reliable values of y_i, with smaller variances σ_i^2, are given more weight. This is achieved by defining a *weighted mean*

$$\overline{y}_w \quad = \quad \sum_{i=1}^{n} v_i y_i, \tag{8.13}$$

where each v_i is a normalised weight which is inversely proportional to the variance of that data point,

$$v_i \quad = \quad \frac{1/\sigma_i^2}{\sum_{j=1}^{n} 1/\sigma_j^2}. \tag{8.14}$$

The normalised weights sum to 1, $\sum_{i=1}^{n} v_i = 1$. As in Equation 3.11, the total error is the sum of two subsidiary error terms,

$$y_i - \overline{y}_w \quad = \quad (y_i - \hat{y}_i) + (\hat{y}_i - \overline{y}_w), \tag{8.15}$$

where $(y_i - \hat{y}_i)$ is the part of the total error *not* explained by the model and $(\hat{y}_i - \overline{y}_w)$ is the part of the total error that *is* explained by the model. The total sum of squared errors is therefore also the sum of two subsidiary sums of squared errors,

$$\sum_{i=1}^{n}(y_i - \overline{y}_w)^2 \quad = \quad \sum_{i=1}^{n}(y_i - \hat{y}_i)^2 + \sum_{i=1}^{n}(\hat{y}_i - \overline{y}_w)^2, \tag{8.16}$$

i.e. the sum of squared errors not explained by the model plus the sum of squared errors that is explained by the model.

However, if we take account of the fact that different data points have different variances then these become weighted sums,

$$\sum_{i=1}^{n}(y_i - \overline{y}_w)^2/\sigma_i^2 = \sum_{i=1}^{n}(y_i - \hat{y}_i)^2/\sigma_i^2 + \sum_{i=1}^{n}(\hat{y}_i - \overline{y}_w)^2/\sigma_i^2, \tag{8.17}$$

i.e. the *weighted* total sum of squared errors is the *weighted* sum of squared errors not explained by the model plus the *weighted* sum of squared errors explained by the model. Recalling that $1/\sigma_i^2$ is the *i*th diagonal element W_{ii} of the matrix W (Equaiton 8.5), this can be written as

$$\sum_{i=1}^{n}(y_i - \overline{y}_w)^2\, W_{ii} = \sum_{i=1}^{n}(y_i - \hat{y}_i)^2\, W_{ii} + \sum_{i=1}^{n}(\hat{y}_i - \overline{y}_w)^2\, W_{ii}. \tag{8.18}$$

Hence the total sum of squared errors can be written in vector–matrix format as

$$SS_T \quad = \quad (\mathbf{y} - \bar{\mathbf{y}}_w)^\mathsf{T} W (\mathbf{y} - \bar{\mathbf{y}}_w) \tag{8.19}$$

$$= \quad (\mathbf{y} - \hat{\mathbf{y}})^\mathsf{T} W (\mathbf{y} - \hat{\mathbf{y}}) + (\hat{\mathbf{y}} - \bar{\mathbf{y}}_w)^\mathsf{T} W (\hat{\mathbf{y}} - \bar{\mathbf{y}}_w), \tag{8.20}$$

where $\mathbf{y}$ and $\hat{\mathbf{y}}$ are the vectors defined in Equations 7.17 and 7.12, and $\bar{\mathbf{y}}_w$ is the column vector whose elements are all equal to $\bar{y}_w$ in Equation 8.13. As in Chapter 3, we define the second term in Equation 8.20, the weighted sum of squares explained by the model, as

$$SS_{\mathrm{Exp}} \quad = \quad (\hat{\mathbf{y}} - \bar{\mathbf{y}}_w)^\mathsf{T} W (\hat{\mathbf{y}} - \bar{\mathbf{y}}_w), \tag{8.21}$$

and we define the first term in Equation 8.20, the noise or residual sum of squares not explained by the model, as

$$SS_{\mathrm{Noise}} \quad = \quad (\mathbf{y} - \hat{\mathbf{y}})^\mathsf{T} W (\mathbf{y} - \hat{\mathbf{y}}), \tag{8.22}$$

(this is the same as E in Equation 8.6), so that Equation 8.20 becomes

$$SS_T \quad = \quad SS_{\mathrm{Exp}} + SS_{\mathrm{Noise}} \tag{8.23}$$

(the same as Equation 3.21). Using these definitions, the proportion of the total sum of squared errors explained by the regression model is

$$r_w^2 \quad = \quad \frac{SS_{\mathrm{Exp}}}{SS_T}, \tag{8.24}$$

and substituting Equations 8.19 and 8.22 yields

$$r_w^2 \quad = \quad \frac{(\hat{\mathbf{y}} - \bar{\mathbf{y}}_w)^\mathsf{T} W (\hat{\mathbf{y}} - \bar{\mathbf{y}}_w)}{(\mathbf{y} - \bar{\mathbf{y}}_w)^\mathsf{T} W (\mathbf{y} - \bar{\mathbf{y}}_w)}. \tag{8.25}$$

Given that $SS_{\mathrm{Exp}} = SS_T - SS_{\mathrm{Noise}}$, Equation 8.24 becomes

$$r_w^2 \quad = \quad \frac{SS_T - SS_{\mathrm{Noise}}}{SS_T} \tag{8.26}$$

$$= \quad 1 - \frac{SS_{\mathrm{Noise}}}{SS_T}, \tag{8.27}$$

and substituting Equations 8.19 and 8.22,

$$r_w^2 = 1 - \frac{(\mathbf{y} - \hat{\mathbf{y}})^{\mathsf{T}} W (\mathbf{y} - \hat{\mathbf{y}})}{(\mathbf{y} - \bar{\mathbf{y}}_w)^{\mathsf{T}} W (\mathbf{y} - \bar{\mathbf{y}}_w)}, \tag{8.28}$$

which can be substituted into Equation 8.12 to obtain an F-ratio. Taking account of the degrees of freedom yields the adjusted r^2,

$$r_{w,\text{Adj}}^2 = 1 - \frac{SS_{\text{Noise}}/(n - p)}{SS_{\text{T}}/(n - 1)}. \tag{8.29}$$

Statistical Significance of Individual Parameters. As in previous chapters, to test the significance of each parameter, we first need to estimate the underlying standard deviation of that parameter. For example, from Equation 7.32, the slope parameter b_1 is associated with a t-value of

$$t_{b1}(\nu) = b_1/\hat{\sigma}_{b1}, \tag{8.30}$$

where $\nu = n - p$, with $p = k + 1$ being the number of parameters for k regressors. Similarly, the intercept parameter b_0 is associated with a t-value of

$$t_{b0}(\nu) = b_0/\hat{\sigma}_{b0}. \tag{8.31}$$

Each of the standard deviations $\hat{\sigma}_{b1}$ and $\hat{\sigma}_{b0}$ is the square root of one diagonal element of the covariance matrix

$$\hat{\sigma}_X^2 = \hat{\sigma}_\eta^2 (X^{\mathsf{T}} W X)^{-1} \tag{8.32}$$

$$= \begin{pmatrix} \hat{\sigma}_{b1}^2 & 0 \\ 0 & \hat{\sigma}_{b0}^2 \end{pmatrix} \tag{8.33}$$

(the analogue of Equation 8.32 in the weighted regression case), where the noise variance σ_η^2 is estimated as

$$\hat{\sigma}_\eta^2 = \frac{E}{n - p}. \tag{8.34}$$

Thus Equation 8.32 can be computed as

$$\hat{\sigma}_X^2 = \left(\frac{E}{n-p}\right)(X^\mathsf{T} W X)^{-1}. \qquad (8.35)$$

8.5. Numerical Example

Assessing the Overall Model Fit. The statistical significance of the correlation coefficient is assessed using the F-ratio in Equation 8.12 (repeated here),

$$F(p-1, n-p) = \frac{r_w^2/(p-1)}{(1-r_w^2)/(n-p)}. \qquad (8.36)$$

From Equations 8.24, 8.21 and 8.19, the coefficient of determination is

$$r_w^2 = \frac{SS_{\text{Exp}}}{SS_{\text{T}}} = \frac{56.390}{124.746} = 0.452. \qquad (8.37)$$

Taking account of the number of parameters yields the adjusted r^2 statistic from Equation 8.29,

$$r_{w,\text{Adj}}^2 = 1 - \frac{1/(n-p)\sum_{i=1}^{n}(y_i - \hat{y}_i)^2}{1/(n-1)\sum_{i=1}^{n}(y_i - \overline{y}_w)^2} \qquad (8.38)$$

$$= 0.402. \qquad (8.39)$$

Substituting $r_w^2 = 0.452$, $p - 1 = 2 - 1 = 1$ and $n - p = 13 - 2 = 11$ into Equation 8.36, we get

$$F(1, 11) = \frac{0.452/1}{(1 - 0.452)/11} = 9.075. \qquad (8.40)$$

	parameter value	standard error $\hat{\sigma}$	t	ν	p
slope b_1	1.511	0.502	3.012	11	0.0118
intercept b_0	2.122	0.623	3.405	11	0.0059
r^2	adjusted r^2	F	ν_{Num}	ν_{Den}	p
0.452	0.402	9.075	1	11	0.0118

Table 8.2: Weighted linear regression of the data in Table 8.1.

For the numerator degrees of freedom $p - 1 = 1$ and the denominator degrees of freedom $n - p = 11$, the p-value associated with this F-ratio is $p = 0.0118$, which is slightly less significant than the value $p = 0.0101$ obtained in Equation 5.54 with the same values of y but *unweighted*.

Statistical Significance of the Individual Parameters. As noted in Figure 8.1, for the data in Table 8.1, weighted regression yields a slope of $b_1 = 1.511$ and an intercept of $b_0 = 2.122$. For comparison, simple regression, which assumes that all data points have the same standard deviation, yields $b_1 = 0.764$ and $b_0 = 3.22$.

The t-value associated with the slope b_1 is

$$t_{b1}(\nu) = \frac{b_1}{\hat{\sigma}_{b1}} \tag{8.41}$$

$$= \frac{1.511}{0.502} = 3.012, \tag{8.42}$$

which, with $\nu = n - p = 13 - 2$ degrees of freedom, has a p-value of $p(3.012, 11) = 0.0118$.

The t-value associated with the intercept b_0 is

$$t_{b0}(\nu) = \frac{b_0}{\hat{\sigma}_{b0}} \tag{8.43}$$

$$= \frac{2.122}{0.623} = 3.405, \tag{8.44}$$

which, with $\nu = n - p = 13 - 2$ degrees of freedom, has a p-value of $p(3.405, 11) = 0.0059$.

Chapter 9

Nonlinear Regression

9.1. Introduction

Even though the title of this book is *Linear Regression*, a summary of nonlinear regression is provided for completeness. Nonlinear regression is required if it is suspected that the data cannot be fitted well with a straight line (i.e. linear) model. This could be the case when we know the exact nature of the physical process that determines how the independent and dependent variables are related. For example, at the start of an epidemic, the number y of infected people increases as an exponential function of time x, so the regression model could be expressed as

$$\hat{y} = b_0\, e^{b_1 x}. \tag{9.1}$$

In general, we can write the nonlinear function as $f(x_i, \mathbf{b})$, so the regression problem involves finding parameter values $\mathbf{b} = (b_0, b_1, \dots, b_k)$ that minimise the sum of squared errors

$$E = \sum_{i=1}^{n} (y_i - f(x_i, \mathbf{b}))^2, \tag{9.2}$$

where the observed value y_i is a noisy version of the value $\hat{y}_i$ given by the model,

$$y_i = \hat{y}_i + \eta_i. \tag{9.3}$$

There are two broad classes of models used to fit nonlinear functions to data, as described in the next two sections.

9.2. Polynomial Regression

Suppose we have reason to believe that the n observed values of y can be fitted not by a line but by a quadratic function of the form

$$y_i = b_0 + b_1 x_i + b_2 x_i^2 + \eta_i, \qquad (9.4)$$

where η_i represents noise, so the predicted value of y at x_i is

$$\hat{y}_i = b_0 + b_1 x_i + b_2 x_i^2. \qquad (9.5)$$

For brevity, the regression coefficients can be represented as the vector

$$\mathbf{b} = (b_0, b_1, b_2). \qquad (9.6)$$

Notice that x^2 (and, more generally, any x^m) is linearly related to $\hat{y}$, so (strictly speaking) fitting such an equation to the data is actually a linear regression problem. As in previous chapters, the least squares estimates of the regression coefficients can be obtained by minimising the sum of squared errors

$$E = \sum_{i=1}^{n} (y_i - \hat{y}_i)^2. \qquad (9.7)$$

Notice that Equation 9.5 has the same form as Equation 7.2 used in multivariate regression, the only difference being that each regressor (x_{i1} and x_{i2}) in Equation 7.2 has been replaced by x_i raised to a particular power (x_i^1 and x_i^2) here. Consequently, we can treat the polynomial regression problem of Equation 9.5 as if it were a multivariate regression problem with two regressors, x_i and x_i^2, as shown in Table 9.1.

i	1	2	3	4	5	6	7	8	9	10	11	12	13
y_i	3.34	4.97	4.15	5.40	5.21	4.56	3.69	5.86	4.58	6.94	5.57	5.62	6.87
x_i	1.00	1.25	1.50	1.75	2.00	2.25	2.50	2.75	3.00	3.25	3.50	3.75	4.00
x_i^2	1.00	1.56	2.25	3.06	4.00	5.06	6.25	7.56	9.00	10.56	12.25	14.06	16.00

Table 9.1: Values of the regressors x and x^2 and the dependent variable y.

Naturally, the fit of a polynomial model to n data points improves as the order of the regression function is increased. At one extreme, if the regression function is a first-order polynomial (i.e. a straight line)

$$\hat{y}_i \;=\; b_1 x_i + b_0, \tag{9.8}$$

the sum of squared errors is almost certainly greater than zero. At the other extreme, if the regression function is a polynomial of high enough order k,

$$\hat{y}_i \;=\; b_1 x_i + b_2 x_i^2 + \cdots + b_k x_i^k + b_0, \tag{9.9}$$

then the fitted polynomial will pass through every data point exactly, so that $\hat{y}_i = y_i$ for all $i = 1, \ldots, n$ and hence the sum of squared errors equals zero. However, this does not necessarily mean that the polynomial model provides an overall good fit to the data; we cannot rely on only the sum of squared errors to assess the fit of the model.

In practice, the quality of the fit is assessed using the *extra sum-of-squares method* introduced in Section 7.6. This consists of

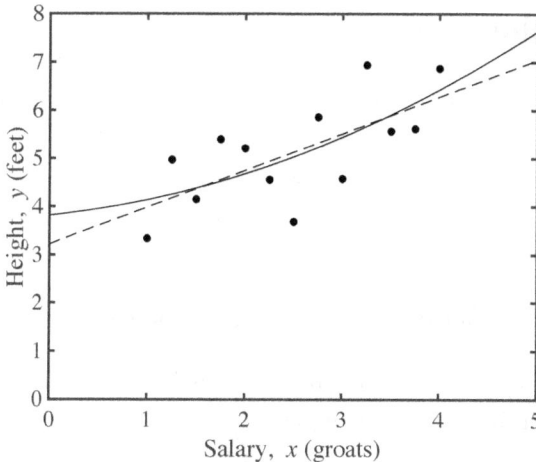

Figure 9.1: The dashed line is the best fitting linear function shown previously ($b_1 = 0.764$, $b_0 = 3.22$; $r_L^2 = 0.466$). The solid curve is the best fitting quadratic function (Equation 9.5 with $b_1 = 0.212$, $b_2 = 0.111$ and $b_0 = 3.819$; $r_{NL}^2 = 0.473$). The improvement from r_L^2 to r_{NL}^2 is assessed using the extra sum-of-squares method (Section 7.6), which yields $p = 0.730$, so the quadratic model does not provide a significantly better fit.

calculating the extra sum-of-squares explained by the model when the polynomial order is incremented by 1, for example from the first-order polynomial in Equation 9.8 to the second-order (quadratic) polynomial in Equation 9.5. If the extra sum-of-squares explained yields an F value that has associated p-value less than 0.05, we reject the null hypothesis that $b_2 = 0$ and conclude that x^2 contributes significantly to the model. A comparison of linear and nonlinear (polynomial) regression is summarised in Figure 9.1.

Multicollinearity

Whereas we could reasonably assume that the regressors x_1 and x_2 in Equation 7.2 are uncorrelated, we know that the regressors x and x^2 in Equation 9.5 are definitely correlated (which can be addressed using *orthogonal polynomials*). This means that we cannot easily assess the statistical significance of the individual coefficients in Equation 9.5. However, we can still assess the statistical significance of the overall fit of the model to the data.

9.3. Nonlinear Regression

If the regression problem cannot be solved using a polynomial function (which, as we have seen, is really a type of multivariate linear regression) then the problem is truly nonlinear. In this case, the regression function falls into one of two categories, as follows.

Regression Functions That Can Be Linearised

Some nonlinear regression problems can be transformed into linear regression problems. For example,

$$\hat{y} = e^{b_1 x} \tag{9.10}$$

can be transformed into a linear regression problem by taking logarithms of both sides,

$$\log \hat{y} = b_1 x. \tag{9.11}$$

Now we view $\log y$ as the dependent variable; standard regression methods assume that its observed values include Gaussian noise η,

$$\log y_i \;=\; b_1 x_i + \eta_i. \tag{9.12}$$

Expressed in terms of the untransformed variable, this is

$$y \;=\; e^{b_1 x + \eta} \;=\; e^{b_1 x} \times e^{\eta}. \tag{9.13}$$

In words, if $\log y$ includes noise η with a Gaussian distribution (as assumed by the regression model) then the noise in y is e^{η}, which has a *log-normal distribution*. This is typically highly skewed, so it violates the Gaussian assumptions on which regression is based.

Regression Functions That Cannot Be Linearised

If the regression function we want to fit cannot be linearised with a transformation then we have two options.

First, we can attempt to approximate the regression function with a polynomial by using its *Maclaurin expansion* (i.e. a Taylor expansion around $x = 0$), so that the method described in Section 9.2 can be applied. For example, if the underlying regression function is known to be exponential, it can be approximated as

$$e^{b_1 x} \;\approx\; 1 + b_1 x + \frac{b_1^2 x^2}{2!} + \frac{b_1^3 x^3}{3!} + \cdots , \tag{9.14}$$

where the polynomial on the right is the Maclaurin expansion of $e^{b_1 x}$. Note that the approximation becomes exact as the number of terms approaches infinity.

Second, we can attempt to find, by brute force, parameter values that minimise the sum of squared errors in Equation 9.2. If the number of parameters is small then such a brute-force method could be exhaustive search as in Section 2.2. However, for a large number of parameters, some form of gradient descent is required, as in Section 2.3.

9.4. Numerical Example

Finding the Best Fitting Plane. Using the data in Table 9.1, the best fitting quadratic curve (Equation 9.5) is shown in Figure 9.1. The least squares estimates of the regression coefficients are

$$b_1 = 0.212,$$
$$b_2 = 0.111,$$
$$b_0 = 3.819.$$

Substituting these values into Equation 9.5, we have

$$\hat{y}_i = 0.212 \times x_i + 0.111 \times x_i^2 + 3.819. \qquad (9.15)$$

Assessing the Overall Model Fit. The square of the correlation coefficient (i.e. the coefficient of determination, Equation 7.28) is

$$r^2 = \frac{\text{var}(\hat{y})}{\text{var}(y)} = \frac{0.518}{1.095} = 0.473, \qquad (9.16)$$

so the correlation coefficient is $r = \sqrt{0.473} = 0.688$.

The statistical significance of the multiple correlation coefficient is assessed using the F-ratio (see Section 5.6). Using Equation 7.31 (repeated below), the F-ratio of the multiple correlation coefficient with numerator degrees of freedom $p-1 = 3-1 = 2$ and denominator degrees

	parameter value	standard error $\hat{\sigma}$	t	ν	p
coefficient b_1	0.212	1.570	2.121	10	0.060
coefficient b_2	0.111	0.310	0.135	10	0.895
intercept b_0	3.819	1.800	0.357	10	0.729
r^2	adjusted r^2	F	ν_{Num}	ν_{Den}	p
0.473	0.368	4.49	2	10	0.041

Table 9.2: Results of polynomial regression analysis of the data in Table 9.1.

of freedom $n - p = 13 - 3 = 10$ is

$$F(p - 1, n - p) = \frac{r^2/(p-1)}{(1-r^2)/(n-p)} \tag{9.17}$$

$$= \frac{0.473/2}{(1-0.473)/10} \tag{9.18}$$

$$= 4.49, \tag{9.19}$$

which corresponds to a p-value of

$$p = 0.041. \tag{9.20}$$

This is less than 0.05, so the multiple correlation coefficient (which represents the overall fit) is statistically significant.

Using the extra sum-of-squares method to assess if $b_2 = 0$. To test the null hypothesis that the regressor x^2 does not contribute significantly to the model, we need to assess the probability that the coefficient $b_2 = 0$. This involves calculating the two sum of squared errors in Equation 7.44,

$$SS_{\text{Exp}}(b_2|\mathbf{b}_{\text{RED}}) = SS_{\text{Exp}}(\mathbf{b}_{\text{FULL}}) - SS_{\text{Exp}}(\mathbf{b}_{\text{RED}}) \tag{9.21}$$

$$= 6.738 - 6.643 \tag{9.22}$$

$$= 0.095. \tag{9.23}$$

So Equation 7.45 (repeated here),

$$F(\nu_{\text{Diff}}, n - p) = \frac{SS_{\text{Exp}}(b_2|\mathbf{b}_{\text{RED}})/\nu_{\text{Diff}}}{SS_{\text{Noise}}/(n-p)}, \tag{9.24}$$

becomes

$$F(1, 10) = \frac{0.095/1}{7.502/10} \tag{9.25}$$

$$= 0.127, \tag{9.26}$$

which corresponds to a p-value of $p = 0.729$. This is greater than 0.05, so we cannot reject the hypothesis that $b_2 = 0$. In other words, the quadratic term x^2 with coefficient b_2 does not contribute significantly to the model.

Chapter 10

Bayesian Regression: A Summary

10.1. Introduction

Bayesian analysis is a rigorous framework for interpreting evidence in the context of previous experience or knowledge. At its core is *Bayes' theorem*, also known as *Bayes' rule*, which was first formulated by Thomas Bayes (c. 1701–1761), and also independently by Pierre-Simon Laplace (1749–1827). After more than two centuries of controversy, during which Bayesian methods have been both praised and pilloried, Bayesian analysis has now emerged as a powerful tool with a wide range of applications, including in artificial intelligence, genetics, linguistics, image processing, brain imaging, cosmology and epidemiology.

Bayesian inference is not guaranteed to provide the correct answer. Rather, it provides the probability that each of a number of alternative answers is true, and these can then be used to find the answer that is most probably true. In other words, it provides an informed guess. While this may not sound like much, it is far from random guessing. Indeed, it can be shown that no other procedure can provide a better guess, so Bayesian inference can be justifiably interpreted as the output of a perfect guessing machine, or a perfect inference engine. This perfect inference engine is fallible, but it is provably less fallible than any other means of inference.

Up to this point, the best fitting line has been defined as the line whose slope and intercept parameters are the least squares estimates (LSE). However, as explained in Chapter 6, the LSE can also be obtained using maximum likelihood estimation (MLE), which provides parameter values that maximise the probability of the data. Of course, what we really want is a method which provides parameter values that are the most probable ones given the data we have. This brings us to a vital, fundamental distinction between two frameworks: the *frequentist framework* and the *Bayesian framework*.

In essence, whereas frequentist methods (like those used in the previous chapters) answer questions regarding the probability of the data, Bayesian methods answer questions regarding the probability of a particular hypothesis. In the context of regression, frequentist methods estimate the probability of the data if the slope were zero (null hypothesis), whereas Bayesian methods estimate the probability of any given slope based on the data, and can therefore estimate the most probable slope. This apparently insignificant difference represents a fundamental shift in perspective.

Subjective Priors. A common criticism of the Bayesian framework is that it relies on *prior distributions*, which are often called *subjective priors*. However, there is no reason in principle why priors should not be objective. Indeed, the objective nature of Bayesian priors can be guaranteed mathematically, via the use of *reference priors*.

A Guarantee. Before we continue, we should reassure ourselves about the status of Bayes' theorem: *Bayes' theorem is not a matter of conjecture*. By definition, a theorem is a mathematical statement that has been proved to be true. A thorough treatment of Bayesian analysis requires mathematical techniques beyond the scope of this introductory text. Accordingly, the following pages present only a brief, qualitative summary of the Bayesian framework.

10.2. Bayes' Theorem

Bayes' theorem, also known as Bayes' rule, says that

$$p(x|y) = \frac{p(y|x)\,p(x)}{p(y)}. \qquad (10.1)$$

Each term in Bayes' rule has its own name: $p(x|y)$ is the probability of x given y, or the *posterior probability*; $p(y|x)$ is the probability of y given x, or the *likelihood* of x; $p(x)$ is the *prior probability* of x; and $p(y)$, the probability of y, is the evidence or *marginal likelihood*. In practice, the results of Bayesian and frequentist methods can be similar because the influence of the prior becomes negligible for large data sets.

Bayesian Regression

Because Equation 10.1 represents a general truth, we can replace x with the regression parameters $\mathbf{b} = (b_0, b_1)$ and y with the n observed values y_i collected in a vector $\mathbf{y}$ as in Equation 6.12, which yields

$$p(\mathbf{b}|\mathbf{y}) \quad = \quad \frac{p(\mathbf{y}|\mathbf{b})\,p(\mathbf{b})}{p(\mathbf{y})}, \tag{10.2}$$

where the likelihood $p(\mathbf{y}|\mathbf{b})$ was defined in Section 6.2.

The difference between the likelihood $p(\mathbf{y}|\mathbf{b})$ and the posterior $p(\mathbf{b}|\mathbf{y})$ looks almost trivial; after all, it involves simply swapping the order in which two terms are written. But the implications of this small difference can be profound. For example, suppose the slope that maximises the likelihood $p(\mathbf{y}|\mathbf{b})$ is $b_{\text{MLE}} = 0.2$. Now suppose a huge of amount of previous experience with the type of data under consideration informs us that the slopes have a narrow Gaussian distribution with a mean of $\bar{b}_1 = 0.5$. Should this prior knowledge affect the final estimate of the slope? Of course it should. And Bayesian analysis tells us (amongst other things) precisely how much to change our estimate of the slope to take account of such prior knowledge.

A Rational Basis For Bias. Naturally, all decisions regarding the value of an estimated parameter should be based on evidence (data), but the best decisions should also be based on previous experience. Bayes' theorem provides a way of taking into account previous experience in interpreting evidence. For example, the precise way in which prior experience affects the decision regarding the best fitting slope is through multiplication of the likelihood for each putative slope by the value of the prior at that slope. This ensures that values of the slope that were encountered more often in the past receive a boost. One might think that this is undesirable because it biases outcomes towards those that have been obtained in the past. But the particular form of bias obtained with Bayes' rule is fundamentally rational. Bayes' theorem is, simply put, a rational basis for bias.

Appendix A

Glossary

alternative hypothesis The working hypothesis, for example that the slope of the best fitting line is not equal to zero. See null hypothesis.

average Usually understood to be the average of a sample taken from a parent population, while the word mean is reserved for the average of the parent population.

Bayesian analysis Statistical analysis based on Bayes' theorem.

Bayes' theorem The posterior probability of x given y is $p(x|y) = p(y|x)p(x)/p(y)$, where $p(y|x)$ is the likelihood and $p(x)$ is the prior probability of x.

chi-squared test (or χ^2-test) Commonly known as a goodness-of-fit test, this is less accurate than the F-test for regression analysis.

coefficient of determination The proportion of variance in a variable y that is accounted for by a regression model, $r^2 = \text{var}(\hat{y})/\text{var}(y)$.

confidence limit The 95% confidence limits of a sample mean $\bar{y}$ are $\mu \pm 1.96\hat{\sigma}_{\bar{y}}$ where μ is the population mean and $\hat{\sigma}_{\bar{y}}$ is the standard error.

correlation A normalised measure of the linear inter-dependence of two variables, which ranges between $r = -1$ and $r = +1$.

covariance An unnormalised measure of the linear inter-dependence of two variables x and y, which varies with the magnitudes of x and y.

degrees of freedom The number of ways in which a set of values is free to vary, given certain constraints (imposed by the mean, for example).

frequentist statistics The conventional framework of statistical analysis used in this book. Compare with Bayesian analysis.

Gaussian distribution (or normal distribution) A bell-shaped curve defined by two parameters, the mean μ and variance σ^2. A shorthand way of writing that a variable y has a Gaussian function is $y \sim \mathcal{N}(\mu, \sigma_y^2)$.

heteroscedasticity The assumption that noise variances may not all be the same.

homoscedasticity The assumption that all noise variances are the same.

inference Using data to infer the value, or distribution, of a parameter.

likelihood The conditional probability $p(y|b_1)$ of observing the data value y given a parameter value b_1 is called the likelihood of b_1.

maximum likelihood estimate (MLE) Given the data y, the value $b_{1\,\text{MLE}}$ of a parameter b_1 that maximises the likelihood function $p(y|b_1)$ is the maximum likelihood estimate of the true value of b_1.

mean The mean of a set of n values of y is $\bar{y} = \frac{1}{n}\sum_{i=1}^{n} y_i$.

multiple correlation coefficient If the proportion of variance in y that is explained by a multivariate regression model is r^2 then the multiple correlation coefficient is r.

multivariate regression A model $\hat{y} = b_0 + b_1 x_1 + b_2 x_2 + \cdots + b_k x_k$ that assumes y depends on k regressors, for which the k regression coefficients and the intercept are to be estimated.

noise The part of a measured quantity that is not predicted by a model.

normal distribution See Gaussian distribution.

null hypothesis The hypothesis that we want to show is improbable given the data (e.g. that the slope of the regression line is zero); precisely how improbable is the p-value.

one-tailed test A test which estimates either the probability that a value $\bar{y}$ is larger (but not smaller) than a hypothetical level $\bar{y}_0$, or the probability that $\bar{y}$ is smaller (but not larger) than $\bar{y}_0$.

p-value The probability that the absolute value of a parameter b is equal to or greater than the observed value $|b_{\text{obs}}|$, assuming that the true value of b is zero. It is a measure of how improbable b_{obs} is, assuming that there is no underlying effect (e.g. slope $= 0$).

parameter A coefficient in an equation, such as the slope b_1 of a line, which acts as a model for observed data.

parent population An infinitely large set of values of a quantity, from which each finite sample of n values is assumed to be drawn.

regressor An independent variable in a regression model; for example, in the model $\hat{y} = b_1 x_1 + b_2 x_2 + b_0$, the regressors are x_1 and x_2, which account for a proportion of the variance in the dependent variable y.

regression A technique used to fit a parametric model (e.g. a straight line) to a set of data points.

sample A set of n values assumed to be chosen at random from a parent population of values.

statistical significance If the probability p that a value arose by chance, given the null hypothesis, is $p < 0.05$ then it is statistically significant.

standard deviation A measure of 'spread' in the values of a variable; the square root of the variance.

theorem A mathematical statement that has been proved to be true.

two-tailed test A test which estimates the probability that an observed value $\bar{y}$ is larger or smaller than (but not equal to) a hypothetical value.

variance A measure of how 'spread out' the values of a variable are.

vector An ordered list of numbers. See Appendix C.

z-score The distance between an observed value y and the mean μ of a Gaussian distribution measured in units of standard deviations.

Appendix B

Mathematical Symbols

$\propto$ proportional to. For example, $y \propto x$ means $y = cx$ where c is a constant.

$\sum$ (capital Greek letter sigma) shorthand for summation. For example, if we have $n = 3$ numbers $x_1 = 2, x_2 = 5$ and $x_3 = 7$ then their sum can be represented as

$$\sum_{i=1}^{n} x_i = 2 + 5 + 7 = 14.$$

The variable i is counted up from 1 to n, and for each i the term x_i adopts a new value, which is added to a running total.

$\prod$ (capital Greek letter pi) shorthand for multiplication. For example, the product of the numbers defined above can be represented as

$$\prod_{i=1}^{n} x_i = 2 \times 5 \times 7 = 70. \tag{B.1}$$

The variable i is counted up from 1 to n, and for each i the term x_i adopts a new value, which is multiplied by a running total.

$\approx$ approximately equal to.

$\geq$ greater than or equal to.

$\leq$ less than or equal to.

ν (Greek letter nu, pronounced 'new') the number of degrees of freedom.

μ (Greek letter mu, pronounced 'mew') the population mean.

η (Greek letter eta, pronounced 'eater') the noise in y.

ε noise, difference between a value y_i and the population mean μ.

σ (Greek letter sigma) the population standard deviation.

$\hat{\sigma}$ unbiased estimate of the population standard deviation based on ν degrees of freedom.

Mathematical Symbols

σ^2 population variance.

$\hat{\sigma}^2$ unbiased estimate of the population variance based on ν degrees of freedom.

b_0 intercept of a line (i.e. the value of y at $x = 0$).

b_1 slope of a line (i.e. the amount of change in y per unit increase in x).

$\text{cov}(x, y)$ covariance of x and y, based on n pairs of values.

E sum of squared differences between the model and the data.

$\overline{E}$ mean squared error; the sum of squared differences divided by the number of data points n ($\overline{E} = E/n$).

k number of regressors in a regression model, which excludes the intercept b_0.

n number of observations in a sample (data set).

p number of parameters in a regression model, which includes k regressors plus the intercept b_0 (so $p = k + 1$). Also p-value.

$r(x, y)$ correlation between x and y based on n pairs of values.

r^2 proportion of variance in data y accounted for by a regression model.

r_w^2 proportion of variance in data y accounted for by a regression model when each data point has its own variance.

s_x standard deviation of x based on a sample of n values.

s_y standard deviation of y based on a sample of n values.

SS_{Exp} (explained) sum of squared differences between the model-predicted values $\hat{y}$ and the mean $\overline{y}$.

SS_{Noise} (noise, or unexplained) sum of squared differences between the model-predicted values $\hat{y}$ and the data y (the same as E).

SS_{T} (total) sum of squared differences between the data y and the mean $\overline{y}$; $SS_{\text{T}} = SS_{\text{Exp}} + SS_{\text{Noise}}$.

var variance based on a sample of n values.

V $n \times n$ covariance matrix in which the ith diagonal element σ_i^2 is the variance of the ith data point y_i.

W weight matrix, $W = V^{-1}$, in which the ith diagonal element is $1/\sigma_i^2$.

$\mathbf{x}$ vector of n observed values of x: $\mathbf{x} = (x_1, x_2, \ldots, x_n)$.

x position along the x-axis.

y position along the y-axis.

$\mathbf{y}$ vector of n observed values of y: $\mathbf{y} = (y_1, y_2, \ldots, y_n)$.

z position along the z-axis. Also z-score.

Appendix C

A Vector and Matrix Tutorial

The single key fact to know about vectors and matrices is that each vector represents a point located in space, and a matrix moves that point to a different location. Everything else is just details.

Vectors. A number, such as 1.234, is known as a *scalar*, and a *vector* is an ordered list of scalars. A vector with two components b_1 and b_2 is written as $\mathbf{b} = (b_1, b_2)$. Note that vectors are printed in bold type.

Adding Vectors. The *vector sum* of two vectors is obtained by adding their corresponding elements. Consider the addition of two pairs of scalars (x_1, x_2) and (b_1, b_2); adding the corresponding elements gives

$$(x_1, x_2) + (b_1, b_2) \;=\; \big((x_1 + b_1), (x_2 + b_2)\big). \qquad (C.1)$$

In vector notation we can write $\mathbf{x} = (x_1, x_2)$ and $\mathbf{b} = (b_1, b_2)$, so that

$$
\begin{aligned}
\mathbf{x} + \mathbf{b} \;&=\; (x_1, x_2) + (b_1, b_2) && (C.2)\\
&=\; \big((x_1 + b_1), (x_2 + b_2)\big) \\
&=\; \mathbf{z}. && (C.3)
\end{aligned}
$$

Subtracting Vectors. Subtracting vectors is done similarly by subtracting corresponding elements, so that

$$
\begin{aligned}
\mathbf{x} - \mathbf{b} \;&=\; (x_1, x_2) - (b_1, b_2) && (C.4)\\
&=\; \big((x_1 - b_1), (x_2 - b_2)\big). && (C.5)
\end{aligned}
$$

Multiplying Vectors. Consider the result of multiplying the corresponding elements of two pairs of scalars (x_1, x_2) and (b_1, b_2) and then adding the two products together:

$$y \;=\; b_1 x_1 + b_2 x_2. \qquad (C.6)$$

Writing (x_1, x_2) and (b_1, b_2) as vectors, we can express y as

$$y \;=\; (x_1, x_2) \cdot (b_1, b_2) = \mathbf{x} \cdot \mathbf{b}, \qquad (C.7)$$

where Equation C.7 is to be interpreted as Equation C.6. This operation of multiplying corresponding vector elements and adding the results is called the *inner*, *scalar* or *dot* product, and is often denoted by a dot, as here.

Row and Column Vectors. Vectors come in two basic flavours, *row vectors* and *column vectors*. The *transpose operator* ⊤ transforms a row vector (x_1, x_2) into a column vector (or vice versa):

$$(x_1, x_2)^\top = \begin{pmatrix} x_1 \\ x_2 \end{pmatrix}. \tag{C.8}$$

The reason for having row and column vectors is that it is often necessary to combine several vectors into a single *matrix*, which is then used to multiply a single column vector $\mathbf{x}$, defined here as

$$\mathbf{x} = (x_1, x_2)^\top. \tag{C.9}$$

In such cases, we need to keep track of which vectors are row vectors and which are column vectors. If we redefine $\mathbf{b}$ as a column vector, $\mathbf{b} = (b_1, b_2)^\top$, then the inner product $\mathbf{b} \cdot \mathbf{x}$ can be written as

$$y = \mathbf{b}^\top \mathbf{x} \tag{C.10}$$

$$= (b_1, b_2) \begin{pmatrix} x_1 \\ x_2 \end{pmatrix} \tag{C.11}$$

$$= b_1 x_1 + b_2 x_2. \tag{C.12}$$

Here, each element of the row vector $\mathbf{b}^\top$ is multiplied by the corresponding element of the column vector $\mathbf{x}$, and the results are summed. This allows us to simultaneously specify many pairs of such products as a vector–matrix product. For example, if the vector variable $\mathbf{x}$ is measured n times then we can represent the measurements as a $2 \times n$ matrix $X = (\mathbf{x}_1, \ldots, \mathbf{x}_n)$. Then, taking the inner product of each column $\mathbf{x}_t$ in X with $\mathbf{b}^\top$ yields n values of the variable y:

$$(y_1, y_2, \ldots, y_n) = (b_1, b_2) \begin{pmatrix} x_{11} & x_{12} & \cdots & x_{1n} \\ x_{21} & x_{22} & \cdots & x_{2n} \end{pmatrix}. \tag{C.13}$$

Here, each (single-element) column y_t is given by the inner product of the corresponding column in X with the row vector $\mathbf{b}^\top$, so that

$$y = \mathbf{b}^\top X.$$

Finally, it is useful to note that $y^\top = (\mathbf{b}^\top X)^\top = X^\top \mathbf{b}$.

Appendix D

Setting Means to Zero

Setting the means of x and y to zero, or *centring* the data, involves defining two new variables

$$
\begin{aligned}
x_i' &= x_i - \bar{x}, & \text{(D.1)} \\
y_i' &= y_i - \bar{y}, & \text{(D.2)}
\end{aligned}
$$

so that the equation of the regression line becomes

$$
\hat{y}_i' = b_1 x_i' + b_0' \qquad \text{(D.3)}
$$

and

$$
y_i' = b_1 x_i' + b_0' + \eta_i', \qquad \text{(D.4)}
$$

where $\bar{x}' = \bar{y}' = b_0' = \bar{\eta}' = 0$ and $\eta_i' = \eta_i$, as shown in Figure D.1.

Proving $\bar{y}'$ Equals Zero. We can check that the new variable y' has a mean of zero:

$$
\begin{aligned}
\bar{y}' &= \frac{1}{n} \sum_{i=1}^{n} y_i' = \frac{1}{n} \sum_{i=1}^{n} (y_i - \bar{y}) & \text{(D.5)} \\
&= \frac{1}{n} \sum_{i=1}^{n} y_i - \frac{1}{n} \sum_{i=1}^{n} \bar{y} & \text{(D.6)} \\
&= \bar{y} - \bar{y} & \text{(D.7)} \\
&= 0. & \text{(D.8)}
\end{aligned}
$$

A similar line of reasoning shows that x' also has a mean of zero.

Proving $\bar{\eta}'$ Equals Zero. For the best fitting slope, the derivative in Equation 2.2 (repeated here in terms of the new variables) equals zero:

$$
\frac{\partial E}{\partial b_0'} = -2 \sum_{i=1}^{n} (y_i' - (b_1 x_i' + b_0')) = 0. \qquad \text{(D.9)}
$$

Since $y_i' - (b_1 x_i' + b_0') = \eta_i'$, we have $\sum_{i=1}^{n} \eta_i' = 0$, so that

$$\overline{\eta}' = \frac{1}{n} \sum_{i=1}^{n} \eta_i' = 0. \tag{D.10}$$

This was also stated in Equation 2.4 for the original variables.

Proving b_0' Equals Zero. Taking the means of all terms in Equation D.4, we have

$$\overline{y}' = b_1 \overline{x}' + b_0' + \overline{\eta}'. \tag{D.11}$$

We know $\overline{x}' = \overline{y}' = \overline{\eta}' = 0$, so $0 = b_1 \times 0 + b_0' + 0$ and hence $b_0' = 0$.

Proving $\overline{\hat{y}}'$ Equals Zero. Taking the means of all terms in Equation D.3, we have

$$\overline{\hat{y}}' = b_1 \overline{x}' + b_0'. \tag{D.12}$$

We know $\overline{x}' = b_0' = 0$, and hence $\overline{\hat{y}}' = 0$.

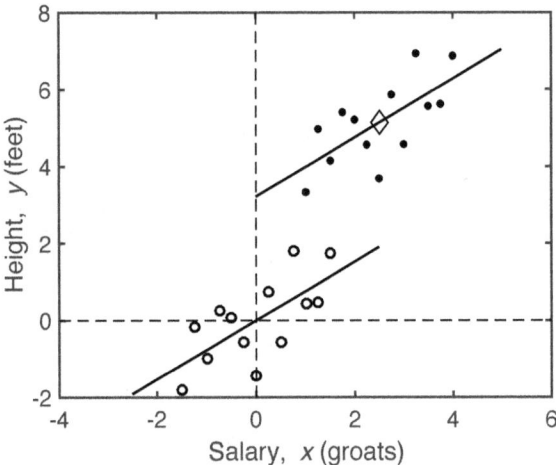

Figure D.1: Setting the means to zero. The upper right dots represent the original data, for which the mean of x is $\overline{x} = 2.50$ and the mean of y is $\overline{y} = 5.13$ (the point $(\overline{x}, \overline{y})$ is marked with a diamond). The lower left circles represent the transformed data, with zero mean. Two new variables x' and y' are obtained by translating x by $\overline{x}$ and translating y by $\overline{y}$, so that $x' = x - \overline{x}$ and $y' = y - \overline{y}$; both x' and y' have mean zero, as indicated by the axes shown as dashed lines. The slope of the best fitting line for regressing y on x is the same as the slope of the best fitting line for regressing y' on x', and for the transformed variables the y'-intercept is at $b_0' = 0$.

Proving $\text{cov}(\hat{y}, \eta) = 0$. Note that translating the data has no effect on the covariance, so that $\text{cov}(\hat{y}, \eta) = \text{cov}(\hat{y}', \eta')$, which is

$$\text{cov}(\hat{y}', \eta') = \frac{1}{n}\sum_{i=1}^{n}(\hat{y}'_i - \overline{\hat{y}}')(\eta'_i - \overline{\eta}'), \qquad (D.13)$$

where $\overline{\hat{y}}' = \overline{\eta}' = 0$, so that

$$\text{cov}(\hat{y}', \eta') = \frac{1}{n}\sum_{i=1}^{n}\hat{y}'_i\,\eta'_i. \qquad (D.14)$$

From Section 2.4, the best fitting line minimises the mean squared error

$$\overline{E} = \frac{1}{n}\sum_{i=1}^{n}(y'_i - \hat{y}'_i)^2, \qquad (D.15)$$

and we know that $\hat{y}'_i = b_1 x'_i$ (because $b'_0 = 0$). At a minimum, the derivative with respect to b_1 is zero:

$$\frac{\partial \overline{E}}{\partial b_1} = \frac{-2}{n}\sum_{i=1}^{n}x'_i(y'_i - \hat{y}'_i) = 0, \qquad (D.16)$$

where

$$(y'_i - \hat{y}'_i) = \eta'_i. \qquad (D.17)$$

So, given that $y'_i = b_1 x'_i + \eta'_i$, we have

$$x'_i = (y'_i - \eta'_i)/b_1 = \hat{y}'_i/b_1. \qquad (D.18)$$

Substituting Equations D.18 and D.17 into Equation D.16 yields

$$\frac{-2}{b_1}\left(\frac{1}{n}\sum_{i=1}^{n}\hat{y}'_i\,\eta'_i\right) = 0. \qquad (D.19)$$

Using Equation D.14 to rewrite the term in brackets, we get $\text{cov}(\hat{y}'_i, \eta'_i) = 0$. Since $\text{cov}(\hat{y}_i, \eta_i) = \text{cov}(\hat{y}'_i, \eta'_i)$, we conclude that

$$\text{cov}(\hat{y}_i, \eta_i) = 0; \qquad (D.20)$$

in other words, the correlation r between $\hat{y}$ and η is zero. From this we know that, $\text{cov}((\hat{y}_i - \overline{y}), \eta_i) = 0$, where $(\hat{y}_i - \overline{y}) = \psi_i$, from which it follows that (as promised in Section 3.4)

$$\frac{1}{n}\sum_{i=1}^{n}\psi_i\eta_i = 0. \qquad (D.21)$$

Appendix E

Key Equations

Variance. Given n values of y_i, the variance is

$$\text{var}(y) \;\; = \;\; \frac{1}{n}\sum_{i=1}^{n}(y_i - \bar{y})^2, \tag{E.1}$$

where $\bar{y}$ is the mean of the y_i values.

Standard deviation. This is the square root of the variance,

$$s_y \;\; = \;\; \left(\frac{1}{n}\sum_{i=1}^{n}(y_i - \bar{y})^2\right)^{1/2}. \tag{E.2}$$

Covariance.

$$\text{cov}(x,y) \;\; = \;\; \frac{1}{n}\sum_{i=1}^{n}(x_i - \bar{x})(y_i - \bar{y}). \tag{E.3}$$

Pearson product-moment correlation coefficient.

$$r \;\; = \;\; \frac{\frac{1}{n}\sum_{i=1}^{n}(x_i - \bar{x})(y_i - \bar{y})}{\left(\frac{1}{n}\sum_{i=1}^{n}(x_i - \bar{x})^2\right)^{1/2}\left(\frac{1}{n}\sum_{i=1}^{n}(y_i - \bar{y})^2\right)^{1/2}}$$

or, equivalently,

$$r \;\; = \;\; \text{cov}(x,y)/(s_x s_y). \tag{E.4}$$

The square of the correlation coefficient, r^2, is called the coefficient of determination, which is the proportion of variance in y accounted for by x; because $r(x,y) = r(y,x)$, r^2 is also the proportion of variance in x accounted for by y.

Regression slope. The least squares estimate of the slope is

$$b_1 = \frac{\frac{1}{n}\sum_{i=1}^{n}(x_i - \bar{x})(y_i - \bar{y})}{\frac{1}{n}\sum_{i=1}^{n}(x_i - \bar{x})^2} = \frac{\text{cov}(x,y)}{\text{var}(x)}. \quad (E.5)$$

The value of t for the difference between b_1 and the null hypothesis value of $b_1' = 0$ is

$$t_{b1} = (b_1 - b_1')/\hat{\sigma}_{b1}, \quad (E.6)$$

where $\hat{\sigma}_{b1}$ is the unbiased estimate of the standard deviation σ_{b1} of b_1,

$$\hat{\sigma}_{b1} = \frac{\left[\frac{1}{n-2}\sum_{i=1}^{n}(y_i - \hat{y}_i)^2\right]^{1/2}}{\left[\sum_{i=1}^{n}(x_i - \bar{x})^2\right]^{1/2}}. \quad (E.7)$$

Here $n-2$ is the number of degrees of freedom. The p-value is associated with t_{b1} and $\nu = n - 2$ degrees of freedom.

p-value for the slope. If the absolute value of t_{b1} is larger than the critical value $t(0.05, \nu)$ then the p-value is $p(t_{b1}, \nu) < 0.05$.

Regression intercept. The least squares estimate of the intercept is

$$b_0 = \bar{y} - b_1\bar{x}. \quad (E.8)$$

The value of t for the difference between b_0 and a hypothetical value b_0' is

$$t_{b0} = (b_0 - b_0')/\hat{\sigma}_{b0}, \quad (E.9)$$

where $\hat{\sigma}_{b0}$ is the unbiased estimate of the standard deviation σ_{b0},

$$\hat{\sigma}_{b0} = \hat{\sigma}_\eta \times \left[\frac{1}{n} + \frac{\bar{x}^2}{\sum_{i=1}^{n}(x_i - \bar{x})^2}\right]^{1/2}, \quad (E.10)$$

where $\hat{\sigma}_\eta$ is defined by Equation 5.13.

p-value for the intercept. If $|t_{b0}| > t(0.05, \nu)$ then the p-value is $p(t_{b0}, \nu) < 0.05$, where the number of degrees of freedom is $\nu = n - 2$.

Statistical evaluation of the overall fit. The statistical significance of the overall fit is assessed using the F-statistic

$$F(p - 1, n - p) = \frac{r^2/(p - 1)}{(1 - r^2)/(n - p)}. \quad (E.11)$$

Index

www.ingramcontent.com/pod-product-compliance
Lightning Source LLC
Chambersburg PA
CBHW071155200326
41519CB00018B/5240